Fish Swimming

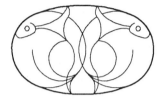

CHAPMAN & HALL FISH AND FISHERIES SERIES

Amongst the fishes, a remarkably wide range of fascinating biological adaptations to diverse habitats has evolved. Moreover, fisheries are of considerable importance in providing human food and economic benefits. Rational exploitation and management of our global stocks of fishes must rely upon a detailed and precise insight of the interaction of fish biology with human activities.

The *Chapman & Hall Fish and Fisheries Series* aims to present authoritative and timely reviews which focus on important and specific aspects of the biology, ecology, taxonomy, physiology, behaviour, management and conservation of fish and fisheries. Each volume will cover a wide but unified field with themes in both pure and applied fish biology. Although volumes will outline and put in perspective current research frontiers, the intention is to provide a synthesis accessible and useful to both experts and non-specialists alike. Consequently, most volumes will be of interest to a broad spectrum of research workers in biology, zoology, ecology and physiology, with an additional aim of the books encompassing themes accessible to non-specialist readers, ranging from undergraduates and postgraduates to those with an interest in industrial and commercial aspects of fish and fisheries.

Applied topics will embrace synopses of fishery issues which will appeal to a wide audience of fishery scientists, aquaculturists, economists, geographers and managers in the fishing industry. The series will also contain practical guides to fishery and analysis methods and global reviews of particular types of fisheries.

Books already published and forthcoming are listed below. The Publisher and Series Editor would be glad to discuss ideas for new volumes in the series . . .

Available titles

1. **Ecology of Teleost Fishes**
   Robert J. Wootton
2. **Cichlid Fishes**
   Behaviour, ecology and evolution
   Edited by Miles A. Keenlyside
3. **Cyprinid Fishes**
   Systematics, biology and exploitation
   Edited by Ian J. Winfield and Joseph S. Nelson
4. **Early Life History of Fish**
   An energetics approach
   Ewa Kamler

5. **Fisheries Acoustics**
   David N. MacLennan and E. John Simmonds
6. **Fish Chemoreception**
   Edited by Toshiaki J. Hara
7. **Behaviour of Teleost Fishes**
   Second edition
   Edited by Tony J. Pitcher
8. **Genetics and Fish Breeding**
   Colin R. Purdom
9. **Fish Ecophysiology**
   J. Cliff Rankin and Frank B. Jensen
10. **Fish Swimming**
    John J. Videler
11. **On the Dynamics of Exploited Fish Populations**
    Raymond J. H. Beverton and Sidney J. Holt
    (Facsimile reprint)
12. **Sea Bass**
    G. Pickett and M. Pawson

Forthcoming titles

Fisheries Ecology
Second edition
Edited by T. J. Pitcher and P. Hart

Hake
Fisheries, products and markets
J. Alheit and T. J. Pitcher

Impact of Species Change in the African Lakes
Edited by T. J. Pitcher

Fish Swimming

JOHN J. VIDELER

Department of Marine Biology
University of Groningen
The Netherlands

CHAPMAN & HALL

London · Weinheim · New York · Tokyo · Melbourne · Madras

Published by Chapman & Hall, 2-6 Boundary Row, London SE1 8HN, UK

Chapman & Hall, 2-6 Boundary Row, London SE1 8HN, UK

Chapman & Hall GmbH, Pappelallee 3, 69469 Weinheim, Germany

Chapman & Hall USA., 115 Fifth Avenue, New York, NY 10003, USA

Chapman & Hall Japan, ITP-Japan, Kyowa Building, 3F, 2-2-1 Hirakawacho, Chiyoda-ku, Tokyo 102, Japan

Chapman & Hall Australia, 102 Dodds Street, South Melbourne, Victoria 3205, Australia

Chapman & Hall India, R. Seshadri, 32 Second Main Road, CIT East, Madras 600 035, India

First edition 1993
Reprinted 1996

Typeset in 10/12 Photina by Acorn Bookwork, Salisbury, Wilts
Printed in Great Britain by St Edmundsbury Press, Bury St Edmunds, Suffolk

ISBN 0 412 40860 0

A Catalogue record for this book is available from the British Library

Library of Congress Cataloging-in-Publication Data available

∞ Printed on permanent acid-free text paper, manufactured in accordance with ANSI/NISO Z39.48-1994 and ANSI/NISO Z39.48-1984 (Permanence of Paper).

Contents

Series foreword

Among the fishes, a remarkably wide range of biological adaptations to diverse habitats has evolved. As well as living in the conventional habitats of lakes, ponds, rivers, rock pools and the open sea, fish have solved the problems of life in deserts, in the deep sea, in the cold antarctic, and in warm waters of high alkalinity or of low oxygen. Along with these adaptations, we find the most impressive specializations of morphology, physiology and behaviour. For example we can marvel at the high-speed swimming of the marlins, sailfish and warm-blooded tunas, air-breathing in catfish and lungfish, parental care in the mouth-brooding cichlids and viviparity in many sharks and toothcarps.

Moreover, fish are of considerable importance to the survival of the human species in the form of nutritious, delicious and diverse food. Rational exploitation and management of our global stocks of fishes must rely upon a detailed and precise insight of their biology.

The *Chapman & Hall Fish and Fisheries Series* aims to present timely volumes reviewing important aspects of fish biology. Most volumes will be of interest to research workers in biology, zoology, ecology and physiology but an additional aim is for the books to be accessible to a wide spectrum of non-specialist readers ranging from undergraduates and postgraduates to those with an intrerest in industrial and commercial aspects of fish and fisheries.

How do fish swim? The short answer is that we still do not know. Fish live with neutral buoyancy in a dense fluid medium, something that is alien to human experience. The ultimate mechanism by which fish gain momentum by imparting force to water remains a mystery and John Videler may surprise many biologists by pointing out that we are as yet unable to apply Newton's laws of motion to fish swimming.

Videler reviews what we know about the essential features about what happens when fish swim through water, from older ideas like levering themselves past pegs on a board, to new discoveries about the role of vortices set up in the water, by the fish's body and tail as it passes, like a trail of footprints. The book thoroughly covers the hardware of fish swimming: the design of swimming muscles and their nervous control, the structure of body skin, fin rays and fins, including appendages modified for propulsion, the cruising muscles well-supplied with oxygen and the emergency acceleration system that enables fish to escape from predators, or be one themselves. This section of the book closes with a discussion of our insight of why there is an optimum shape for fast swimmers and other locomotory specialists.

Kinematics of swimming are dealt with next: how force from the tail is transmitted to the water, how much distance is gained per tail beat, known as the 'stride length'. For those interested in a dynamic perspective of the energetic constraints on fish moving through their environment, the book closes with a novel outline of the costs of swimming in different ways and at different speeds.

This volume in the *Chapman & Hall Fish and Fisheries Series*, beautifully illustrated by Dick Visser, brings together material that has not recently been subjected to comprehensive and unified review in a book. In addition, this work has the virtue of presenting new ideas exemplified by several case studies, putting an innovative perspective on the problems of fish swimming. Videler, who is not afraid of admitting our ignorance in some critical fields of fish locomotion, aims to inspire and help new researchers to ask helpful questions and apply the best techniques. I am confident that this book will help him meet this aim for some years to come.

Preface

WHY IS FISH SWIMMING AN INTERESTING SUBJECT?

The aquatic environment is the cradle of life on earth. After 10^9 years of evolution, the more than 22 000 extant fish species show a rich variety of adaptations to meet the requirements of this environment. Of course, not all the adaptations displayed by fish species are related to swimming, and good swimmers can also be found among other taxonomic groups. Fishes are, however, masters of the art of propulsive interaction with water.

Fish live submerged, surrounded by fluid, where buoyancy balances gravity to make their apparent weight close to zero. Because fish are neutrally buoyant and not in direct contact with the earth, fish swimming is fundamentally different from terrestrial locomotion. From a human point of view, the aquatic way of life is alien and fish may be less familiar to us than fictitious extra-terrestrials.

Fish swimming is also of interest because of the variety of disciplines involved. The physical laws governing the interactions between fish and water mould and dictate the details of morphological, physiological and behavioural adaptations. These laws and the forces generated for propulsion are described in Chapter 1. Fish, of course, have to obey Newton's laws of motion, but it is not easy to see how they do so. Although knowledge of the hydrodynamic aspects has accumulated over the past two decades, the process of the exchange of momentum, as required by Newton, is not yet fully understood. The main reason for this lack of understanding is that we do not know precisely what a fish does to the water through which it moves. For those readers interested in quantifying thrust forces, Lighthill's Large-Amplitude Elongated-Body theory is explained and used in an example.

Chapters 2, 3 and 4 deal with the functional morphology of the swimming apparatus. Chapter 2 considers muscles and nerves. Among a large number of anatomical adaptations, the lateral muscles on both sides of the median septum are probably the most exclusive features of a fish. The short lateral muscle fibres, closely packed in metameric myotomes, occupy virtually the whole body behind the head and around the abdominal cavity. The myotomes are not simply straight blocks of muscles, staggered one behind the other; their shape is spatially complex, forming double cones in forward and backward directions. The reason for this complexity is still largely unknown. Fish axial muscle fibre structures and the innervation of these fibres deviate from the normal vertebrate configuration for reasons we sometimes do, but often do not, understand.

Chapter 3, on body axis and fin structure and function, shows how the median septum is supported by the vertebral column to make it into a flexible shaft, incompressible in its longitudinal direction. Fins are designed to transmit remotely the forces exerted by the muscles inside the fish and to apply these appropriately to generate thrust. Teleost fin rays are structurally and mechanically the most complex, to meet these demands. Assembled into paired and unpaired fins, fin rays form the basic structures to perform a variety of locomotory functions.

Chapter 4 reveals the good reasons why the bodies of fast swimmers should be streamlined, and why there is an optimum shape, offering the largest volume for the lowest drag.

Knowledge of the mechanical properties of the skin and scales is essential if we are to understand how the swimming apparatus works. Ideas about the function of peculiar external anatomical features, such as the sword of swordfish and the sharp horizontal keels on the caudal peduncle of fast swimmers, are also included in Chapter 4. A variety of adaptations diversifies fish species into groups of locomotory specialists.

Fish movements have always astonished human observers. Chapter 5 offers a brief historical overview of the progress made in attempts to describe the kinematic details of the movements scientifically. Descriptions of different swimming styles resulted in complex classifications and definitions in the beginning of the 20th Century. We will not use these, but will instead concentrate on analysing techniques to study a variety of movements in Chapter 6. The vast majority of fish use lateral undulations of the body and tail for propulsion at all speeds. Most of the rest use paired or unpaired fins, and not the body, to swim at lower speeds; they use their body only at high speeds. During steady swimming, movements are rhythmic, and the distance moved forward for every swimming stroke is regarded as the stride length of a fish. Its maximum value appears to differ slightly between species. The stride length is usually constant during steady swimming and does not change with speed. The speed is therefore determined by the stride frequency. Fish usually do not swim in a steady fashion: rapid starts, turning, swimming with periodic bursts of activity followed by coasting with a stiff, straight body and braking are the most common manoeuvres. Detailed studies of the movements of fins greatly improved the knowledge of the complex interactions between fish and water.

Chapter 7 explores our knowledge of fish muscle physiology, starting at the level where the forces originate inside the muscle fibres. Phenomena determining the magnitude of the forces and the dynamics of force and work generation are described. Fish muscle fibres usually operate dynamically: during rhythmic sinusoidal changes in length, stimulation may occur at different instants during the length-change cycle. The total work done per cycle, which equals the force times the displacement, can be graphically represented by a work loop emerging when force is plotted against change in length. The surface area of the loop represents the total work done. This technique has been applied to find optimal relations between innervation patterns and length-change cycles.

The fastest contractions of lateral muscles determine the maximum speed of fish. These have been measured *in vitro*, using electrically stimulated blocks of muscle fibres. Results show a strong positive correlation with temperature and a decrease with increasing body size. These maximum frequency relations can be used to predict maximum speed once the stride length of a species is known.

Patterns of muscle activities, used by the fish to generate the complex movements, have been studied with electromyographic techniques.

Chapter 8 describes a mathematical model of the forces and bending moments of undulating fish in interaction with reactive forces from the water. These forces delay the speed at which the propulsive wave of curvature runs down the fish body. The speed of the wave of curvature is lower than that of the wave of contraction of the lateral muscles from head to tail. In some species the muscles can even be simultaneously active along one side of the body during steady swimming. Simultaneous studies of electromyograms and kinematics reveal that model predictions are realistic. These findings show consequently that fish use their lateral muscles differently along the body: the muscles in the front part provide most of the work and the rearward muscles transmit high forces while resisting stretching. Speculations about transmission of forces from the main bulk of muscles to the tail blade generated an unconventional idea about the transfer of force by muscles. We, as terrestrial animals, are used to muscles that generate force by trying to bring the two ends of the muscle closer together. We think of muscles as shortening machines. Body-builders, however, show off their strength by demonstrating how their muscles swell out when activated. Since muscles occupy a constant volume, the bulging of a muscle could generate its force in a direction around its circumference instead of in the longitudinal direction. All we need is a transducer for the force in that direction. The tight fish skin, with its crisscross layers of collagenous fibres, is probably such a transducer.

Chapter 9 compares costs of swimming at the cheapest speeds. Many fish species spend most of their time hardly swimming at all, showing only small displacements within a restricted area, such as a territory or a feeding range. This behaviour reminds us of the fact that the metabolic cost of swimming increases approximately with the cube of the speed. There is an optimum speed where the amount of energy per unit distance covered reaches a minimum. This is likely to be the rather slow speed at which fish travel most of the time.

Oxygen consumption has been measured to obtain an estimate of the energy expenditure during swimming. The energy needed to transport one unit of weight over one body length at optimum speed is considered the fairest, dimensionless basis for a comparison between animals of a wide range of sizes. The data available for larval and adult fish are compared with the swimming costs at the optimum speed of other submerged and surface swimmers. Surface swimmers face high drag penalties owing to wave formation. Optimal-speed swimming costs of fish are compared with those of surface and submerged swimmers, with the costs of flying at maximum-range speeds of animals and aircraft, and with the costs of running of terrestrial animals.

Ecological implications of the costs of swimming at various speeds and the limits of speed and endurance in fish are discussed in Chapter 10. Limits are of course important survival factors for a fish during escape or attack, but endurance is, although less spectacular, at least equally important. In fish, endurance is not limited at speeds below the maximum sustained speed. It rapidly declines above that speed and is reduced to a fraction of a minute at the maximum prolonged speed; it is reduced to a few seconds at maximum burst speeds. Factors shown to influence maximum swimming performance are discussed.

Swimming costs may make up a substantial part of the energy budget of an individual fish. Ultimately, the allocation of time spent swimming at various speeds should be translated into the allocation of energy, as part of the energy budgets of individual fish measured in the field, to provide insight into the importance of swimming in relation to growth and reproduction.

In summary, its challenging complexity makes fish swimming an interesting subject. This complexity includes basic physics and specific hydrodynamic principles, macroscopic and microscopic anatomical details, novel approaches in muscle physiology, kinematics of steady and unsteady movements, measurements and estimates of energy expenditure and ecological aspects. Study of each of these aspects leads to the frontiers of knowledge. Surveying the relationships between the factors involved in fish locomotion generates feelings of satisfaction and disappointment, satisfaction because of the tremendous body of knowledge amassed over the last century and disappointment about the large gaps in our knowledge. In fact we still do not know how fish swim.

WHAT KIND OF BOOK IS THIS AND WHO ARE THE EXPECTED READERS?

The flow of scientific papers has increased dramatically in recent decades. It is almost impossible to keep up with the literature if one's interests are expanded to more than one narrowly defined subject. One important purpose of scientific books is to give an overview of the state of the art in a multidisciplinary field. This book tries to meet this purpose and therefore includes detailed knowledge of physics (in particular fluid mechanics), functional anatomy, muscle physiology and biomechanics, kinematics, energetics and ecological aspects.

The reader will probably notice that this is a rather unconventional scientific book. It deals with general trends as usual, but also explores the frontiers of scientific knowledge using specific examples in detail and discusses some methods that are used to tackle particular questions. The reason for this dualistic approach is that I tried not only to give an overview and explanation of established knowledge and common facts, but also to update the reader with current discussions and research interests. The choice of topics is of course biased by my own interests, but the aim was to reflect emphasis in the field. I attempted not only to give results but also to provide insight into how experiments are done by describing a number of case studies. Owing to the subjects treated, Chapters 8 and 10 probably have an unevenly large share of the case study approach.

In some instances a rather detailed mathematical treatment is used, including a fair number of equations. Each equation is also explained in detail in the text and the reader should be able to follow the arguments without studying the formulae.

The book is aimed at students interested in animal locomotion. Those who want to specialize in aquatic locomotion will use it as a starting point for their own research. It should help them to ask the right questions and to apply the right techniques. Courses in biomechanics or functional morphology could use it as a source of information and of supplementary reading. Zoology teachers may want to update their lecture notes and could find many data compiled in easy-to-read figures explaining a variety of subjects, from propulsive force generation by moving fish tails through fin ray structure and function to examples of positive and negative work loops in muscles. This book is intended to inspire hydrodynamicists, functional morphologists, physiologists and ecologists alike.

Acknowledgements

Although this book has a single author, it is the result of the efforts of a large group of extremely co-operative relatives, colleagues and friends. The series editor Dr Tony Pitcher generated the idea and convinced me that I should do it. Two esteemed colleagues, Dr Paul Webb and Dr Clem Wardle, commented on all the chapters and greatly influenced the final version. Sir James Lighthill explained his large-amplitude elongated-body theory to me in great detail and gave me the confidence to write the simplified version of it at the end of Chapter 1. Dr Felix Hess and Dr Charlie Ellington helped to improve Chapter 1. The expert advice I needed for the chapters on muscles was provided by Drs Rie Akster, Ian Johnston, Willem van Raamsdonk and Roger Woledge. Dr Peter Geerlink reviewed the chapters on functional morphology.

My wife Hanneke acted as my co-worker; she collected and computerized about a thousand references, read through the various versions of each chapter and insisted on clarity. Not a single chapter would have been written without her invaluable talent for organization and order.

At the start of the two-year period during which we worked on the book, my research group allowed me to spend five sabbatical months in the United Kingdom. Without this flying start it would have taken much longer. We enjoyed the splendid hospitality of Hans and Jane Kruuk in Aboyne while working at the Marine Laboratory in Aberdeen, Neill and Ann McNeill Alexander made us feel at home in Leeds, and Charlie and Stephanie Ellington helped to make our three months' visit to Cambridge a pleasant one. The hospitality of the SOAFD Marine Laboratory in Aberdeen, the Department of Pure and Applied Biology in Leeds and the Department of Zoology of the University of Cambridge is gratefully acknowledged.

Dick Visser is responsible for the artwork. He redrew all the existing figures and designed the original ones. He very often had to force me to be clear and precise.

I thank my research group for their keen interest and patience.

The unremitting support of my daughters Tennie and Hetteke and their interest in my scientific work greatly encourages me.

I thank Chapman & Hall's Nigel Balmforth for his faith in my abilities. The publishing efforts of Martin Tribe are gratefully acknowledged. I am especially indepted to the copy-editor, Charles Hollingworth, who encouraged and supported this venture from start to finish. His interest in the subject, and relentless precision, greatly improved the accuracy of this book.

The errors and omissions are my responsibility. I shall be grateful to anyone who will take the trouble to point them out to me.

Chapter one

Interactions between fish and water

1.1 INTRODUCTION

The physical properties of water and the hydrostatic and hydrodynamic interactions between fish and water are the subject of this chapter. The word 'interactions' puts emphasis on what happens at the interface between the two entities.

Density is the most important physical property in relation to static and dynamic interactions with water. Density-related inertial forces are dominant in fish propulsion. Although viscous forces are small, viscosity is important for the flow very close to the fish and is a main source of drag during swimming. Very small fish larvae have to deal with relatively large viscous forces. Archimedes' law, Newton's principles and Bernoulli's equation govern the magnitude and direction of the forces exchanged between a moving fish and its direct environment.

The starting point for our understanding of fish locomotion is the visualization of what happens when a fish disturbs a static mass of water by moving through it. Two attempts are then made to explain, in broad terms, how fish propel themselves. The first concentrateson what a fish does to the water through which it swims. The second shows how movements of body and tail of a typical fish provoke a reactive force from the water, propelling the fish.

1.2 PROPERTIES OF WATER

Water is an almost incompressible fluid; the density (ρ_w) or weight of fresh water is about 1000 kg m^{-3} at 1 atm and 15 °C, with a viscosity (μ) of about

0.0011 Pa s. Viscosity may be thought of as the internal friction of a fluid. Its unit, the Pascal second (1Pa s = 1 N s m^{-2}) is that of force times distance divided by area times velocity. Both density and viscosity decrease with increased temperatures. They increase when the water becomes more saline, reaching values of 1025 kg m^{-3} and 0.0012 Pa s for sea water of 35 ‰ at 1 atm and 15 °C. The relative importance of density and viscosity varies for fish of different sizes and with swimming speed. It can be estimated by looking at the orders of magnitude of the related forces. Bernoulli's equation (page 6) shows how density-related inertial forces are in the order of $\rho_w u^2 L^2$. Stoke's law predicts the viscous forces to be in the order of $\mu u L$, where L is the size of the animal in the direction of movement in m, u the speed in m s^{-1}. For a 0.3 m fish swimming at 3 body lengths s^{-1} (L s^{-1}) inertial forces are about 36 N and viscous forces only about 0.0003 N, a difference in the order of 10^5 times. This difference is reduced to about 10 times for a 3 mm larva swimming at the same relative speed. The Reynolds number (*Re*) conveniently expresses the relative importance of inertial over viscous forces in a dimensionless way:

$$Re = u \, L \, \rho_w \, \mu^{-1} \tag{1.1}$$

The ratio of dynamic viscosity over density is termed the kinematic viscosity. The reciprocal of the kinematic viscosity at 1 atm and 15 °C is 8.8×10^5 s m^{-2} for fresh water and 8.5×10^5 s m^{-2} for sea water of 35 ‰ salinity.

Re increases from less than 1 for protozoa to 10^8 for the blue whale (Fig. 1.1 and Table 1.1).

Animals in the lower range, including fish larvae, belong to the plankton. They are too small and weak to swim independently of the water currents, so

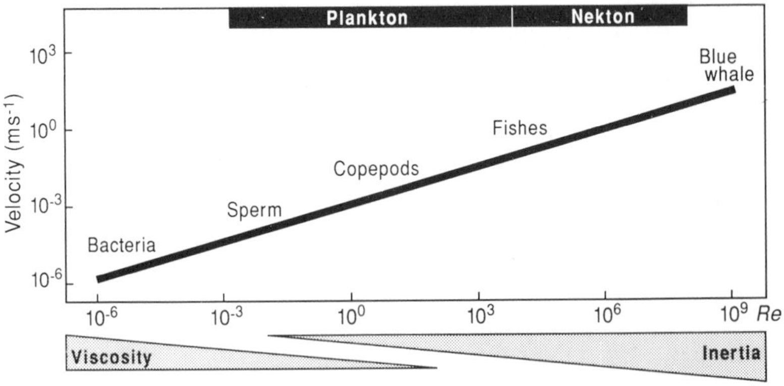

Fig. 1.1 Typical velocities of the largest possible variety of swimmers as a function of *Re*.

Table 1.1 Typical swimming speeds, approximate sizes and matching *Re* numbers of the largest possible variety of swimming organisms

Species	*Velocity* $(m\ s^{-1})$	*Length* (m)	*Re*
Blue whale	10	30	3×10^8
Tuna	10	3	3×10^7
Human	1.7	1.8	3×10^6
Mackerel	3.3	0.3	1×10^6
Herring			
Adult	1	0.2	2×10^5
Larvae	0.5	0.1	5×10^4
	0.16	0.04	6×10^3
	0.06	0.02	1×10^3
	0.02	0.01	2×10^2
Copepods	0.002	0.001	2
Sea urchin sperm	0.0002	0.00015	3×10^{-2}

their main concern is to avoid sinking, and weight and buoyancy are the important forces. At high Reynolds numbers, swimmers are larger than plankton and powerful enough to travel independently of the water movements. Fish, whales, dolphins and squid belong to this group of animals, called nekton. The horizontal thrust and drag forces are dominant, and many adaptations are directed at drag reduction and/or the enhancement of thrust. Anatomical details, representing some of these adaptations in fish are to be found in Chapter 4.

1.3 BUOYANCY, WEIGHT, THRUST AND DRAG

Underwater, the weight of an animal is counteracted by its buoyancy. Archimedes' law tells us that buoyancy (A, in N) equals the weight of the displaced mass of water:

$$A = V \rho_w g \tag{1.2}$$

where g is the acceleration due to gravity and V the volume of the animal.

The weight (W, in N) of the animal is described by

$$W = V \rho_a g \tag{1.3}$$

where ρ_a is the average density of the animal. The apparent weight, $W - A$(N), will usually be close to zero in pelagic fishes. The distribution of the volume along the body determines the position of the centre of buoyancy, whereas relative densities of different parts of the body determine where the resultant weight force acts on the body.

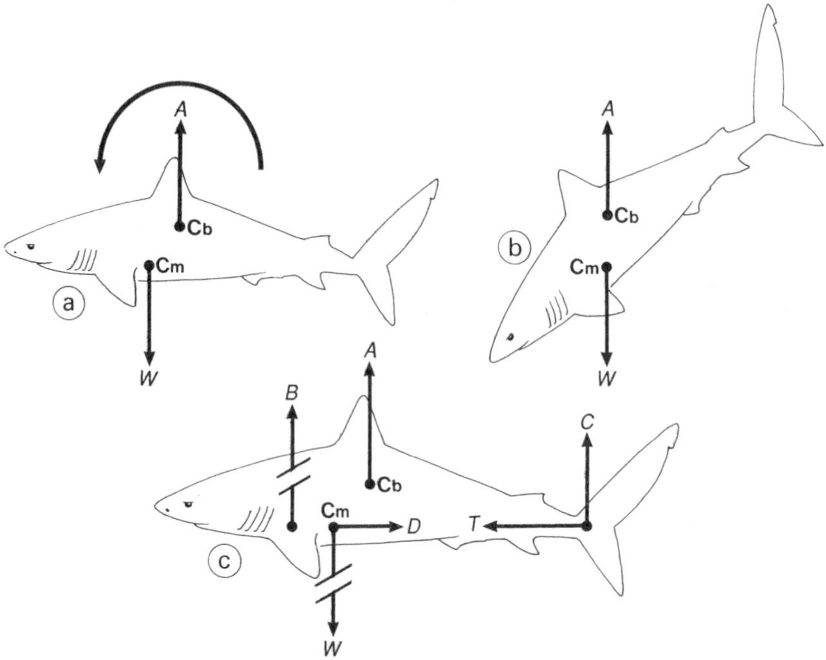

Fig. 1.2 (a) In this resting shark the centre of buoyancy (Cb) and the centre of mass (Cm) are not situated on one vertical line. Buoyancy (*A*) and weight (*W*) will therefore tend to rotate the animal anti-clockwise. (b) The rotation ends when the equilibrium posture is reached where Cb and Cm are on the same sounding-line. (c) The resultants of horizontal and vertical forces on a swimming shark. Equilibrium requires that the sum of the horizontal forces and the sum of the vertical forces be zero, and that the clockwise and anti-clockwise moments of rotation eliminate each other. See text for explanation of forces *B*, *C*, *D* and *T*.

The resting shark in Fig. 1.2(a) is unstable because the centre of buoyancy (Cb) and centre of mass (Cm) are not in the same position along the body. Buoyancy and weight tend to rotate the body anti-clockwise until they act along the same line and Cb is exactly above Cm (Fig. 1.2(b)). The difference in magnitude between *A* and *W* determines whether the fish sinks to the bottom or rises to the surface in the head-down equilibrium posture. Fish use gas-filled swim bladders or materials less dense than water (sharks use oily livers) to lower their average density and thus improve buoyancy. In many species Cm is situated dorsally of Cb so the stable equilibrium position of these fish is belly-up. Being unstable may have an advantage because it provides a higher degree of instant manoeuvrability.

Forces on a swimming animal can be resolved to the ones drawn in Fig.1.2(c). These are more complicated than the resultant forces in the resting situation. Horizontal forces T and D in the median plane are the resultants of all the thrust and drag forces; both were made to act along a horizontal line through the centre of mass. The rear part of the body generates the thrust with lateral movements of the tail. In the case of the shark depicted here, the dorsoventral asymmetry of the tail blade generates also a force C in an upward direction. The extended pectoral fins of the shark act as the wings of an aircraft: they make a slight angle with the horizontal plane and generate upward forces B. The equilibrium conditions for the forces (in N) are:

$$T = D \tag{1.4}$$

$$A + B + C = W \tag{1.5}$$

and for the moments (in N m) of these forces:

$$A\,a + C\,c = B\,b \tag{1.6}$$

where a, b and c are the perpendicular distances between the lines of action of the forces A, B and C and the line of action of W. (T and D could of course also produce rotational moments, if they would not act through the centre of mass.)

The equations of motion and Newton's principles

Before we continue, we have to define precisely what motions or movements actually are. Kinematics is the science of movements without reference to forces. Frame-by-frame analysis of a film of a moving fish taken with a camera in a fixed position shows that the fish is in a different place with respect to the background on every frame. The displacement dx in the swimming direction X, during the time interval dt between the frames can be measured. The velocity (u, in $\mathrm{m\,s^{-1}}$) of the fish is now precisely defined as the rate of change of the displacement.

$$u = \mathrm{d}x/\mathrm{d}t \tag{1.7}$$

If the fish was not moving at a uniform or constant speed, its acceleration and deceleration (a and $-a$, in $\mathrm{m\,s^{-2}}$) are the rates of change of velocity:

$$a = \mathrm{d}u/\mathrm{d}t \tag{1.8}$$

Newton (1686) described the basic principles of movement in his famous *Philosophiae Naturalis Principia Mathematica*, as the three laws of motion. The movements of fish obey these laws and we have to study them if we want to understand fish swimming. The original text of the first law is: '*Corpus omne perseverare in statu suo quiescendi vel movendi uniformiter in directum, nisi quatenus a viribus impressis cogitur statum illum mutare.*' Motte (1729) was the first to translate this into English as: 'Every body perseveres in its state of rest or of uniform motion in a right line, unless it is compelled to change that state by forces impress'd thereon.' This law describes an equilibrium that applies to a non-moving fish or to a fish swimming at a constant speed with thrust T equal to drag D. However, fish are rarely seen to swim at constant speeds; they usually accelerate or decelerate because thrust and drag are not balanced. In this case, Newton's second law applies. The wordings of the second and third laws have been slightly modified over the ages. I have changed the texts used in modern handbooks on physics into fish-related terms. The second law can be paraphrased: 'The rate at which the velocity of a fish changes is equal to the resultant of all the forces on the fish divided by its body mass (M). The direction of the rate of change of velocity is the direction of the resultant force.' If we apply this law to forces (in N) in the swimming direction, we find the equation of motion of an accelerating fish:

$$T - D = M\,a \tag{1.9}$$

The third law describes the exchange of forces between water and fish: 'Every action of the fish on the water will be opposed by an equal reaction in opposite direction.' The actions or forces used for swimming are of course the key issues in a book on fish swimming.

Bernoulli's equation

When an incompressible fluid flows through a horizontal tube of varying cross section, the velocity changes according to a fixed pattern. In narrow parts the flow will be faster than in the wide parts of the tube. The pressures in the tube are measured to be high in parts where the velocity is slow and low in the flow that is accelerated through the narrow sections. Bernoulli derived (in 1738) an equation describing the relation between velocity and pressure in a fluid when viscosity effects can be neglected. The sum of the static pressure p and the dynamic pressure, $0.5\,\rho_w\,U^2$ (both in $\mathrm{N\,m^{-2}}$) is constant in a steady flow without rotation:

$$0.5\rho_w U^2 + p = \text{constant} \tag{1.10}$$

where U is the velocity and ρ_w the density of the fluid. The dynamic pressure is caused by colliding fluid particles.

A swimming fish is much more complicated than a horizontal tube of varying cross section. The fish itself varies in cross section; this variation will cause changes in velocity even along a steadily gliding fish. An actively swimming fish causes far more complex movements of the water and hence we have to deal with a complex pattern of pressure differences.

1.4 THE FOOTPRINTS OF A FISH

Let us look at the fish from the point of view of the water with its inertial and viscous properties. This fluid resists the movements of the fish but simultaneously also enables its propulsion. The available visualizations of the flow around a swimming fish are not detailed enough to show how static water precisely reacts when a fish penetrates and passes through.

First, I will summarize and annotate published flow visualization results and interpret these with the use of basic hydrodynamic principles. A coasting fish, gliding with a straight body, offers the simplest interaction. Then, I shall use the limited data available to reconstruct the impact of lateral oscillating swimming movements of the rear part of the body and tail on the water.

Flow visualization techniques

Hydrodynamicists usually study flow around an object by mounting it in a static position in a steady flow of water. Particles or dye in the water are used to visualize the flow pattern near the model object. A set-up where an object is moving through a stationary fluid is considered physically identical, and is used in flow visualization experiments with swimming fish.

Rosen (1959) made a 4 cm long pearl danio, *Brachydanio albolineatus*, swim over a 2 mm thick layer of milk spread over the bottom of the tank, and filmed the disturbances caused by the passing fish. McCutchen (1976) used layers of water of different temperatures and hence different refractive indices, to make the wake of another species of danio visible. The zebra danio, *Brachydanio rerio*, (length 3.15 cm) was swimming near the thermocline, mixing the water of the two temperatures; the wake of mixed waters was made visible as a shadowgraph. Aleyev (1977) injected an azurine-stained paste into the oral cavity of at least nine species of fish and made them swim in a large tank past a film camera. The blue dye coloured the expired water flowing from the gill clefts into the wake around the fish.

None of these approaches provides a complete picture of what really happens. Rosen's pictures represent the flow underneath a fish, probably distorted by the close presence of the bottom of the tank. In McCutchen's case the exact position of the fish with respect to the thermocline is unknown, which makes it difficult to interpret the pictures. Aleyev made only a part of the interaction between fish and water visible, and assumed that the exhalant water mixes without creating disturbance in the flow around the body. However, his results are probably the best we have. None of the techniques shows the trajectories or streamlines of water particles. However, Aleyev's pictures, in combination with some basic knowledge of hydrodynamics (Prandtl and Tietjens, 1934), will provide a reasonably close representation of what is really happening.

The flow around a coasting fish

To avoid undue complexity, we don't want to start with an actively swimming fish. Instead we study the case where a fish stops its swimming movements after reaching a certain speed. It straightens its body and coasts along a straight line, decelerating slowly. Thrust is no longer produced, so the equation of motion (Equation 1.9) is reduced to:

$$D = M\,a \tag{1.11}$$

Imagine the head of this fish entering undisturbed water as depicted in Fig. 1.3, based on the coasting annular bream, *Diplodus annularis*, filmed by Aleyev. Viscosity causes a layer of water particles very close to the fish to travel with it at the speed of the fish. There is a gradient of decreasing water

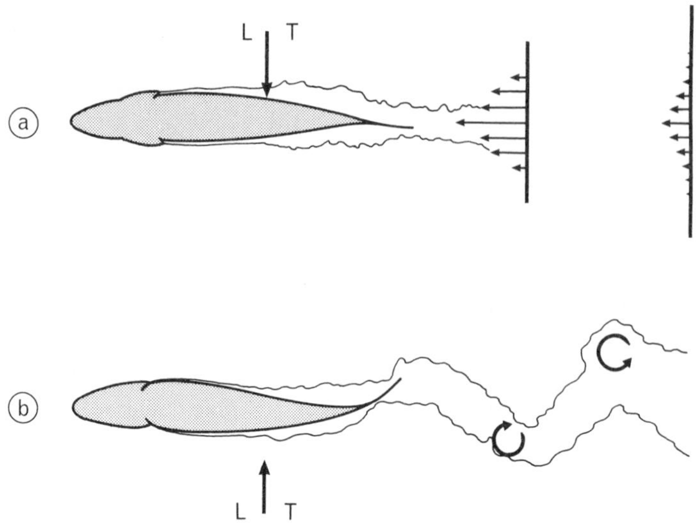

Fig. 1.3 Drawings of the flow patterns induced by a coasting (a) and swimming (b) *Diplodus annularis*, viewed from above, based on pictures by Aleyev (1977). The flow was partly made visible using dye squirting out of the gill clefts of the fish. (a) The flow around a coasting fish changes from laminar to turbulent at a position slightly more than half-way down the body. At two distances behind the fish, estimated forward velocity profiles of water particles in the wake are indicated. (b) Visualization of the flow pattern, caused by a swimming 16.4 cm long fish, shows a transition from laminar (L) to turbulent (T) flow at 10 cm behind the head and stationary vortices in the wake. The swimming speed is 0.76 m s^{-1} and the *Re* at the transition point is 6.5×10^4. The whole fish is swimming at *Re* 1.2×10^5.

velocities in the direction away from the surface of the fish. The viscosity-affected layer of water around the fish is called the boundary layer. Its thickness is defined as the distance from the fish to where the water is no longer dragged along.

From the point of view of the water, the fish, with its water cover, is bumping into stationary water particles and these are accelerated forward and aside. At the rear of the fish, particles instantaneously fill the gap left by the passing fish. Around the fish, moving water particles will transfer some momentum to the neighbouring ones, which will in turn pass it on to particles further away from the fish. Very far away from the fish, the water remains stationary and unaffected by the events.

Equation 1.1 will now be used to calculate a local *Re* value. In this case, the parameter *L* in the equation does not represent the total body length, but the distance along the body from the point of the head to the point where we want to calculate the *Re* number. The *Re* increases with the distance along the fish. The *Re* at 1 mm behind the tip of the head of a fish, coasting at a velocity of, let us assume, 0.5 $\mathrm{m\,s^{-1}}$ in sea water, is about 400. Viscous forces will damp out erratic movements of water particles and these will move along smooth paths or streamlines. This type of flow is called laminar. At 10 cm from the leading edge the *Re* has reached a value of over 40 000, and the viscosity is now so much dominated by the effects of the density that inertial effects become noticeable. The paths of the water particles are no longer smooth, but become erratic. The laminar boundary layer gradually turns into a turbulent one. The dye squirting out of the gill clefts of the fish in Fig. 1.3 gives an indication of the position along the body where this happens. The *Re* will be in the order of 10^5.

Let us try to imagine the fate of water particles in the boundary layer when a fish passes. If a particle happens to be near the surface of the fish, it will travel at the speed of the fish. A particle outside the boundary layer is not dragged along at all. Particles that are initially situated between these two are accelerated in a direction slightly outward from the swimming direction and start to move at an average speed lower than that of the fish. They are slowly overtaken. The wake of the fish is formed by all the particles that are left behind, while still travelling in a direction which is forward on average. Viscosity of the fluid widens the wake and gradually reduces the speed of the particles. The wake will eventually disappear when the particles have transferred their kinetic energy to heat owing to mutual friction.

Flow and the drag force on a coasting fish

We have seen that the water close to the body moves with the velocity of the fish (u) and that water beyond the boundary layer is not dragged along in the swimming direction. This velocity gradient will cause friction between water particles travelling at different speeds. The drag resulting from these processes is called the skin friction drag and is proportional to the surface area (A_w) of the fish.

The fish bumps into static water that has to give way and flows around the fish. This causes congestion in front of the thickest part or shoulder of the fish, resulting in increased pressures there. Rearward of the shoulder the pressure is reduced. These differences in pressure along the body cause the so-called pressure drag. It is proportional to the dynamic pressure, which equals $0.5\,\rho_w\,u^2$ ($N\,m^{-2}$) according to Bernoulli (Equation 1.10). We find both the surface area and the expression for the dynamic pressure in the semi-empirical equation commonly used to describe the drag (in N) on any object including fishes:

$$D = 0.5\,\rho_w\,u^2\,A_w\,C_{dw} \tag{1.12}$$

where C_{dw} is the drag coefficient, a dimensionless number needed to obtain the right magnitude of the drag force. It can be conveniently used to compare drag of different objects at different speeds in fluids of different densities (water and air for example).

An alternative way to derive an equation for the drag force was proposed to me by C.P. Ellington (Department of Zoology, Cambridge University). He suggests that a fish with a frontal area of A_f (m^2), moving at u ($m\,s^{-1}$), will affect a volume of water proportional to $A_f\,u$ ($m^3\ s^{-1}$). This volume multiplied by the density of the water gives the mass of water involved: $\rho_w\ A_f\ u$ ($kg\,s^{-1}$). If this is multiplied by velocity we have the rate of change of momentum, which equals the force (in N), proportional to $\rho_w\ A_f\ u^2$. The equation then is:

$$D = 0.5\,\rho_w\,u^2\,A_f\,C_{df} \tag{1.13}$$

The factor 0.5 has been added by convention. C_{df} represents the proportionality coefficient for this relation and it depends on the shape of the body and on the Reynolds number. (Note that the drag coefficient in this case is a factor A_w/A_f larger than C_{dw}.)

For streamlined bodies such as that of a fish, C_{dw} decreases with increasing *Re* from values in the order of 1 at an *Re* of 100, possibly down to the order of 0.001 at *Re*s above 10^6. Its value is strongly influenced by the state of the flow in the boundary layer. The decrease with increasing *Re* is steeper when the flow is laminar than when it is turbulent.

The drag coefficient can suddenly increase when the flow changes from laminar to turbulent. The drag increases drastically when the water particles fail to follow the contour of the body. In this case the flow separates, usually in the form of circulating masses of water called eddies or vortices. The available flow visualization data indicate that separation does not occur in the boundary layer of a coasting fish.

The drag coefficient of a coasting fish can be calculated from Equations 1.11 and 1.12. A 0.3 m cod, *Gadus morhua*, will serve as an example. Its body mass was 0.3 kg and its surface area, measured from photographs during the same coasting bout, was 0.027 m^2. This piece of coasting started when the cod stopped swimming actively at a speed of 1.23 $m\,s^{-1}$. The animal kept body and tail straight, the paired pectoral and pelvic fins flat against the body and the dorsal and anal fins partly collapsed. It decelerated on average at 0.46 $m\,s^{-2}$ for 1.22 s to a speed of 0.67 $m\,s^{-1}$. The drag coefficient turns out to be 0.011 (Videler, 1981). It will of course be drastically increased as soon as the fish sticks out its fins or bends its body. This happens when a fish brakes (Chapter 6) or swims actively with undulating body and tail.

The flow pattern caused by active swimming

Lateral undulations of the body and the tail blade (described in detail in Chapter 5) deflect the straight path of the wake and produce the backward momentum used to propel the fish. Fig. 1.3(b) is based on a photograph (Aleyev, 1977) of a swimming annular bream, squirting blue dye from its gill clefts. The flow pattern around this 16.4 cm long fish changes from laminar to turbulent at a point on the body at about 10 cm from the head. The speed of the fish was 0.76 $m\,s^{-1}$ and *Re* at the transition point is 6.5×10^4. There is no indication of separation of the flow along the body. In that case the dye would be seen to move away from the outline of the body. The undulating body obviously has influenced the shape of the mass of coloured water as it is left behind. The water surrounding the trailing edge of the tail blade flows off in a direction tangential to the fin. The average direction of the velocity of this water is opposite to the swimming direction, and is contrary to that near the tail of the coasting fish. Every switch in direction of the tail beat causes a change in the direction at which the water is shed, and a rotating movement of the mass of water which is released during that instant. The direction of the rotation was indicated by Aleyev. Looking at it from above, the rotation is clockwise on the left and anti-clockwise on the right of the mean path of swimming. The centres of these rotating masses of water remain approximately in the position where they are shed, and the fish moves away from them. Furthermore we note that the rotating masses are connected by straight traces of dye and that the width of the dye stream and the diameter of the rotating masses increase further down the wake. How can we explain this top view of the three-dimensional 'footprints' of a fish?

1.5 TRANSMISSION OF FORCES BETWEEN FISH AND WATER

The explanation of how forces are produced should ideally start with an outline of the way in which muscle contractions move the body or parts of the body of a fish, and how these movements provoke reactive forces from the water. This sequence of dynamic processes turns out to be very complicated, however, and large parts of it are not yet fully understood. The generation of movements by muscle activity will be the subject of Chapters 7 and 8. The swimming movements are dealt with in Chapter 6. At this point I only want to produce general ideas about how forces are transmitted to the water by swimming fish, rather than to try to quantify these (this will be attempted in Chapter 8).

Fishes typically swim at *Re*s where inertial forces are dominant. The oscillating movements of the body and fins generate momentum proportional to the velocity of the oscillations and the mass of the affected water. According to Newton's third law, the rate of change of momentum in the direction opposite to the swimming direction will result in a thrust force on the fish.

Fish swim by generating a jet of water on average in the direction opposite to the swimming direction. Around a jet of fluid there is a ring of rotating water particles (a vortex ring) situated between the jet and the water at rest around it. The momentum of the jet can either be characterized by the shape and velocities of the vortices or by a description of the jet stream itself.

How each species exactly generates a propulsive jet varies considerably. Fish with lunate or wing-shaped tail blades (e.g. mackerel, *Scomber scombrus*) use the movements and structural properties of the tail fin to generate the propulsive jet. The main body is round on transection and the narrow caudal peduncle, barely interferes with the interaction between oscillating tail and water. I will use this type of propeller to demonstrate and emphasize vortex formation.

The bodies of fish like cod and trout, *Oncorhynchus mykiss*, on the other hand, are laterally compressed, the caudal peduncles are fairly high and the tails have roughly the shape of a flat rectangular plate. In these fish the jet is generated by the pushing action of body and tail, in an obliquely backward direction, and released by the end of the tail blade. The swimming mode of this type of fish is used to concentrate on the accelerated mass of water around the fish that leaves the tail blade as a jet. One of Lighthill's (1975) models is used below to explain the balance of momentum around fish with slender bodies.

Vortex theory of swimming

Aerodynamic theory (Prandtl and Tietjens, 1934) explains how an aerofoil influences the airflow creating vortices to generate lift and thrust. In water, a fish tail acting as a hydrofoil generates thrust using the same principles.

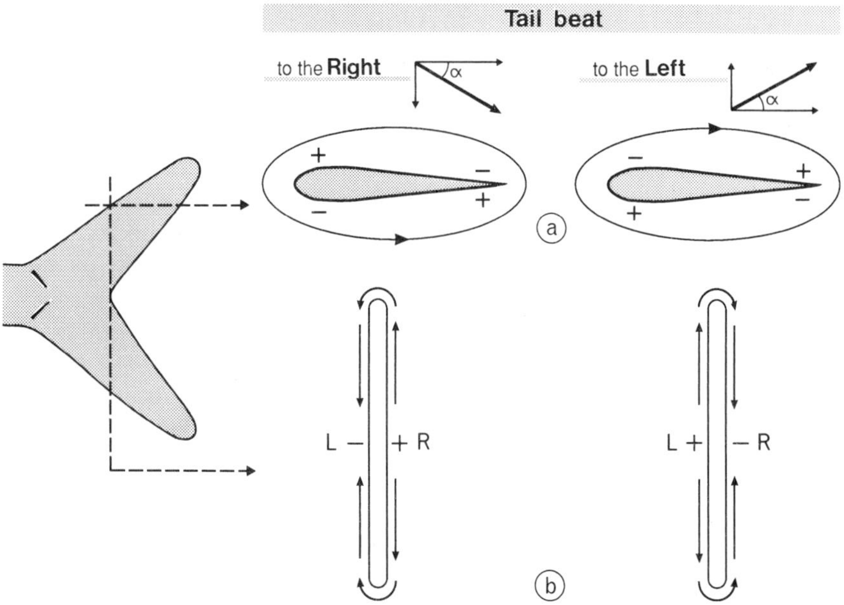

Fig. 1.4 The development of the bound and tail-tip vortices during tail beats to the left and right of the tail of a mackerel. (a) The circulation around a horizontal section during a tail beat to the right and to the left. The thick arrows indicate the direction of the incident flow in each case. This flow is caused by the lateral and forward movement of the tail, indicated by the thin arrows of the inset graphs. α is the angle of incident. (b) Flow circulation and pressure differences between the left (L) and right (R) side of the fin during tail beats to the right and to the left. The vertical sections are viewed from behind the fish.

We take a closer look at the left–right beating tail of a mackerel in Fig. 1.4. The leading edge of the tail fin is rounded and the trailing edge is sharp, as is illustrated by the shape of the horizontal sections (Fig. 1.4(a)). The velocity and angle of attack (α) of the incident flow are determined by the swimming velocity and the lateral velocity of the tail blade. The flow hits the tail fin obliquely, causing stagnation of the water flow and positive pressure on one side, just beyond the leading edge. Consequently, another positive pressure builds up at a rear stagnation point on the opposite side, anterior to the trailing edge. Fast flow and negative pressure exists on the other side of the sharp trailing edge. The flow will now be forced to rotate over a certain angle to reduce the pressure differences. The direction of the rotation will be the one that moves the rear stagnation point in the direction of the trailing edge. The pressure difference reaches a minimum when the rear stagnation point coincides with the trailing edge. (This rotation process is known in fluid dynamics as ‘the Kutta condition’, which says that a body with a sharp trailing edge moving through a fluid will create a circulation about itself of sufficient strength to hold the rear stagnation point at the trailing edge.) Fig. 1.4(a) shows that the direction as observed from above is anti-clockwise during a tail beat to the right, and clockwise during a tail beat to the left. This circulation, which is called the bound vortex, occurs dorsoventrally over the total span of the tail fin. It builds up at the beginning of each lateral tail stroke and keeps its strength until the movement in one direction stops. At that instant the vortex will be left behind and the fin will build up a new one in the opposite direction during the tail beat to the other side.

A two-dimensional view of the wake, in a horizontal plane through the middle of the fish, is drawn in Fig. 1.5(a). It shows how the vortices, released at the end of every stroke to left or right, gradually increase in size. The mutual distances between the cores of the vortices do not change. The gradual decay of the circulation velocity is suggested by the decreasing length of the arrows. Lighthill (1969), after Karman and Burgers (1934), predicts a similar vortex street behind a swimming fish (Fig. 1.5(b)). Note that Lighthill's vortices do not enlarge further downstream because the fluid is assumed to be inviscid. Lighthill suggests the existence of an undulating jet of water threading between the left and right rows of vortices. The strength of the momentum in this jet is equal, and its direction opposite, to the momentum of the fish.

A vertical section through the fin of Fig. 1.4(b) shows the pressure

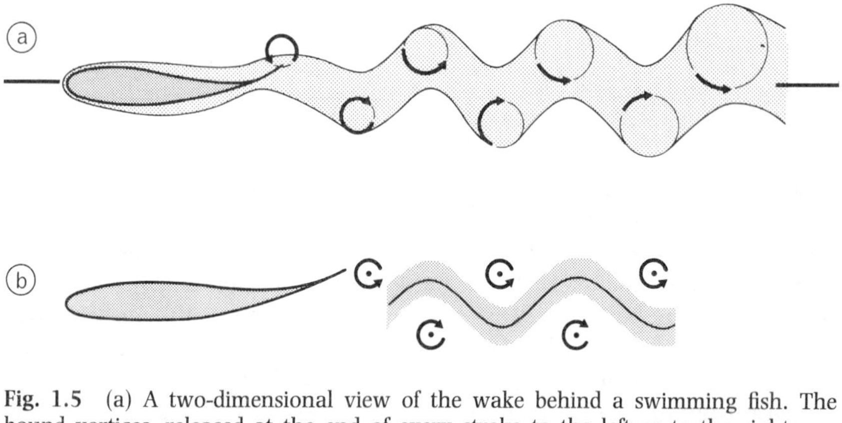

Fig. 1.5 (a) A two-dimensional view of the wake behind a swimming fish. The bound vortices, released at the end of every stroke to the left or to the right, are shown. The arrows indicate the direction of rotation, and their length the relative velocity of the vortices. Note that the vortices remain stationary and that their size gradually increases. (b) Lighthill's (1975) prediction of a vortex street behind a swimming fish under the assumption that viscosity can be entirely neglected.

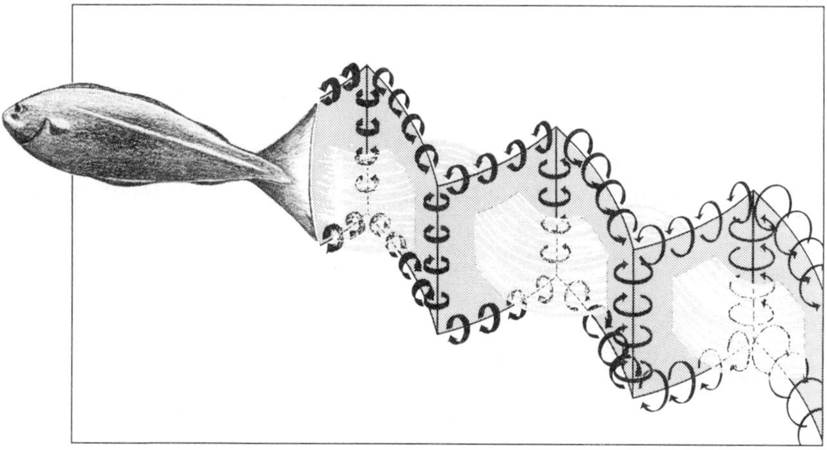

Fig. 1.6 An impression of the three-dimensional vortex flow system and the resulting jetstream behind a saithe, *Pollachius virens* swimming steadily at uniform speed. The tail blade is drawn in the middle of a stroke to the left. The thickness of the arrows indicates relative velocity. Note that the velocity decreases to the right and that the diameter of the vortices increases with increasing distance from the tail.

difference between the leading and the trailing side. At the dorsal and ventral tips of the fin, flow circulating over the edges to the other side tries to compensate for this pressure difference. This rotating flow is left behind as dorsal and ventral trailing vortices during the tail beat from one side to the other. Note that the dorsal and ventral vortices rotate in opposite directions. (These vortices are the equivalent of the wing-tip vortices of aircraft which jet airliners frequently visualize as two white traces while flying at high altitude in a blue sky; condensing exhaust gases are sucked into the wing-tip trailing vortices.)

We are now able to make up a three-dimensional picture. Fig. 1.6 shows how the standing vortices, left behind at the turning points of the tail strokes, are connected by the dorsal and ventral trailing vortices. The jet stream is supported by the surrounding vortices in each successive, slightly bent, rectangle.

The available flow visualizations of vortex sheets provide no indication that the undulating jet exists. Aleyev's picture (Fig. 1.3(b)) is consistent with our three-dimensional drawing if we assume that all the dye has been sucked into the vortices. Precise measurements of the flow around and behind a swimming fish are required to establish how close our three-dimensional picture approximates to reality. Knowledge of dimensions and velocities of the vortex wake would raise the possibility of calculating the forces and energy involved in swimming.

Large-amplitude elongated-body theory

A fish like cod swims with lateral undulations of its body. Precise analyses of steady movements, filmed from above, reveal that the lateral curvature of the body forms a wave running from head to tail with increasing amplitude. (A detailed account of fish swimming kinematics is given in Chapter 6.) For the moment it is sufficient to realize that owing to this movement, every part

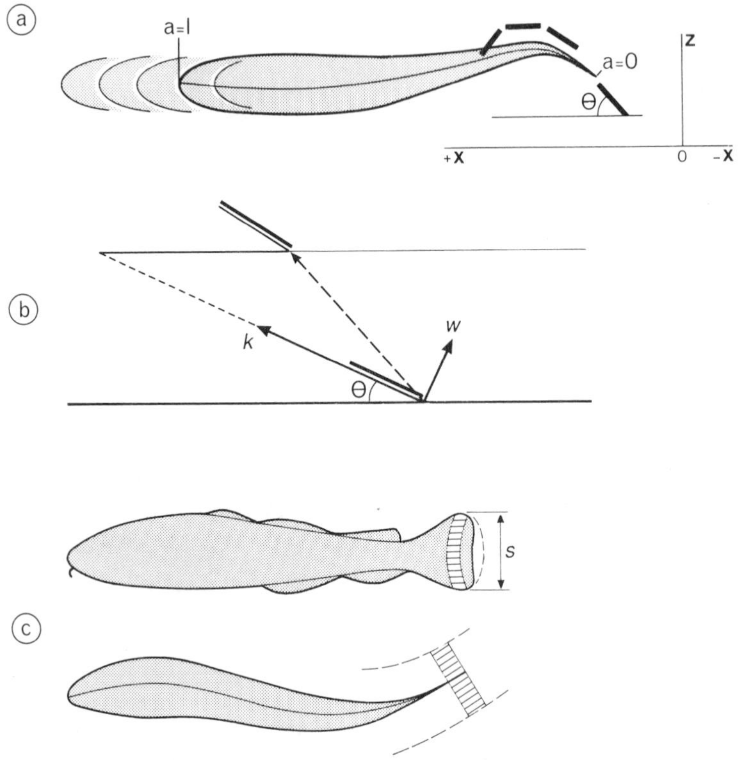

Fig. 1.7 (a) Dorsal view of the outline of a swimming cod drawn from a cine-film. One previous position of head and tail and three subsequent ones indicate the displacement of the fish in time. The swimming direction is along the positive x-axis to the left and z is the perpendicular axis in the horizontal plane. The angle between the end of the tail blade and the x-axis is Θ. (b) Two subsequent positions of the end of the tail blade from Fig. 1.7(a) are redrawn on a larger scale. The broken arrow indicates the instantaneous velocity vector of the tail tip. It has been resolved into a component k in the direction of the tail blade and a component w, perpendicular to k. (c) A lateral and dorsal view of a swimming cod, illustrating the virtual mass of water affected per unit length near the end of the tail blade. The span of the tail is s. The virtual mass of water affected by the lateral tail movements is thought to be restricted to a circular space around the tail end with a radius of $0.5\,s$.

of the body moves forward and oscillates sideways through the water. The average forward velocity u is the same for every part of the body.

The largest lateral amplitude and hence the highest lateral velocity is found at the tail blade. Fig. 1.7 (A) shows a picture of the outline of a steadily swimming cod as seen from above, together with one previous and three subsequent positions of the head and the tail. The swimming direction is along the x-axis, which is positive to the left. The z-axis is the perpendicular axis in the horizontal plane. The centre line of the fish divides the body into lateral halves. A coordinate a describes the centre line between the tail tip where $a = 0$ and the point of the head identified by $a = L$. (L is the length of the body).

From the point of view of the water the sequence of events is as follows: the head of the cod enters undisturbed static water, the water is pushed to the left and right and backward, in the opposite swimming direction, by the passing body with increasing amplitude until the tail blade has passed and no more disturbance will be added. The backward push propels the fish by generating enough thrust to overcome the drag of a fish swimming at uniform speed.

Lighthill's (1971) large-amplitude elongated-body theory describes in detail how the complex fish movements generate thrust and laterally directed forces to both sides. The following description of that theory is restricted to the events in the swimming direction only. Lateral forces are left–right symmetric during straight forward swimming and cancel each other out on average.

The length of the fish is thought to consist of a large number of very small pieces, da long. The interactions between fish and water are principally the same for every part da of the body. To get a rough idea of what is happening between the lateral surface of the fish and the water, we will concentrate for a moment on the events near the end of the tail blade. This is a practical piece of fish to concentrate on, because it is thin, has the largest amplitude (and therefore the largest impact on the water) and the interaction between fish and water ends there. The angle between the surface of the fish and the swimming direction x is Θ. Two successive tail positions from Fig. 1.7 (a) are enlarged and redrawn in Fig. 1.7 (b). The tail tip has moved in the time between the two frames of film from one position to the next as a result of the combined forward and lateral movements. The broken arrow represents the instantaneous velocity vector of the tail tip. This vector is resolved into components k tangential to the blade and w perpendicular to it. The viscosity is assumed to be negligible and therefore movement of the fin in the k-direction causes little disturbance in the water. Vector component w, however, imparts its velocity to a large mass of water on both sides of the fin. Lighthill made an estimate of this added or virtual mass of water by assuming that the movement will affect water passing through a circle around the fish. The part of that circle near the end of the tail is shown in lateral and dorsal view in Fig. 1.7 (c). The affected mass of water per unit length, m (in $\mathrm{kg\,m^{-1}}$), is in that case:

$$m = \rho_w \, \pi \, s^2 \, / \, 4 \tag{1.14}$$

Where s is the span of the tail and $(\pi s^2 / 4)$ the area of the circle around the end of the tail blade. The momentum per unit length $\mathrm{d}a$ of the fish in the opposite swimming direction is $m\ w\ (\text{-}\sin\Theta)$. (The unit of momentum is N s, and momentum per unit length is expressed in $\mathrm{N\,s\,m^{-1}}$, which reduces to $\mathrm{kg\,s^{-1}}$ because $\mathrm{N = kg\,m\,s^{-2}}$. The minus sign denotes the direction of the momentum as the negative x-direction.) The total rate of change of momentum in the opposite swimming direction is described by the rate of change of the integral of the momentum per unit length from head to tail as:

$$\frac{\mathrm{d}}{\mathrm{d}t}\int_0^L mw(\text{-}\sin\Theta)\mathrm{d}a$$

The rate of change of momentum has the unit of force (N). ($\mathrm{kg\,s^{-1}}$ times distance per unit time equals $\mathrm{kg\,m\,s^{-2}}$.) This expression is actually more complicated than it looks, because both m and w vary with the position on the body a. The virtual mass of water m depends on the local body height, and the local velocity perpendicular to the surface of the body w changes with the amplitude of the lateral movement from head to tail.

Lighthill explains how the total rate of change of momentum described by the expression above is, in accordance with Newton's third law, equal to rate of change of momentum in the opposite direction consisting of three components: (1) the reactive force P propelling the fish forward; (2) rate of change of momentum that is left behind at the end of the tail blade; (3) rate of change of momentum due to pressure differences between the part of the water around the fish and the wake behind the fish. To elucidate component (2) we will have to go back to what happens at the end of the tail blade shown in Fig. 1.7 (b). The tail slips with velocity $k\ \mathrm{m\,s^{-1}}$ through the circle, leaving km kg water per second behind. This mass of water moves at speed w in the direction perpendicular to the tail blade (wkm in $\mathrm{kg\,s^{-1} \times m\,s^{-1} = N\,s\,s^{-1} = N}$). The rate of change of momentum, in the opposite swimming direction, left behind in the wake is $mwk\ (\text{-}\sin\Theta)$ where $\text{-}\sin\Theta$ provides the component in the negative x-direction. The velocity vectors k and w can be calculated from the velocities $\mathrm{d}x/\mathrm{d}t$ in the x- and $\mathrm{d}z/\mathrm{d}t$ in the z-direction of the tail tip from:

$$k = \mathrm{d}z/\mathrm{d}t \sin\Theta + \mathrm{d}x/\mathrm{d}t \cos\Theta \tag{1.15}$$

and

$$w = dz/dt \cos \Theta - dx/dt \sin \Theta \tag{1.16}$$

The origin of component (3), the rate of change of momentum due to pressure differences, is less easy to imagine. A pressure difference based on Bernoulli's principle is thought to exist between two sides of an imaginary plane at right angles to the end of the tail blade. The dynamic pressure caused by the tail pushing water at velocity w is according to Bernoulli's equation, $0.5\rho_w w^2\,\mathrm{N\,m^{-2}}$. The total area involved in this pressure force is the area of the circle around the end of the tail blade ($\pi s^2/4$), giving a total pressure force of $0.5mw^2$ N. Lighthill (1970 and personal communication) uses the unsteady form of Bernoulli's equation to prove that the integrated pressure over a plane perpendicular to the end of the tail blade equals w times the momentum mw in the w direction minus the kinetic energy $0.5mw^2$. The total pressure will thus be $mw^2 - 0.5mw^2 = 0.5mw^2$. The same result is obtained by Lighthill (1970) using arguments taken from vortex theory. The component in the swimming direction of this pressure force is $0.5mw^2(-\cos\Theta)$. The minus sign shows that the pressure is pushing backwards in the negative x-direction.

The sum of all the backward-pushing forces and the reactive thrust force P in the positive x-direction must, according to Newton, be equal to zero:

$$\frac{d}{dt}\int_0^L mw(-\sin\theta)da + [mwk(-\sin\theta) + \frac{1}{2}mw^2(-\cos\theta)]_{tail} + P = 0 \tag{1.17}$$

The average value of the integral term of Equation 1.17 is zero because the rate of change of a periodically oscillating quantity is zero. If that is the case, the average thrust force P equals:

$$\overline{P} = \overline{(mwk\sin\theta + \frac{1}{2}mw^2\cos\theta)_{tail}} \tag{1.18}$$

Results of propulsive force estimates with the relevant kinematic parameters are shown in Table 1.2 and Fig. 1.8 for an average tail beat of a 0.30 m cod swimming steadily at a mean velocity of $0.86\ \mathrm{m\,s^{-1}}$. The force increases rapidly after the beginning of each stroke in concert with fast increases in lateral tail tip velocity.

It is good to realize at this stage that some of the assumptions made are not strictly true. Viscous effects cannot be entirely denied, and the mass of water affected does not have a circular shape on transection nor has it a uniform speed. These simplifications are expected to create only small errors.

The average of P over several cycles gives an estimate of the total force

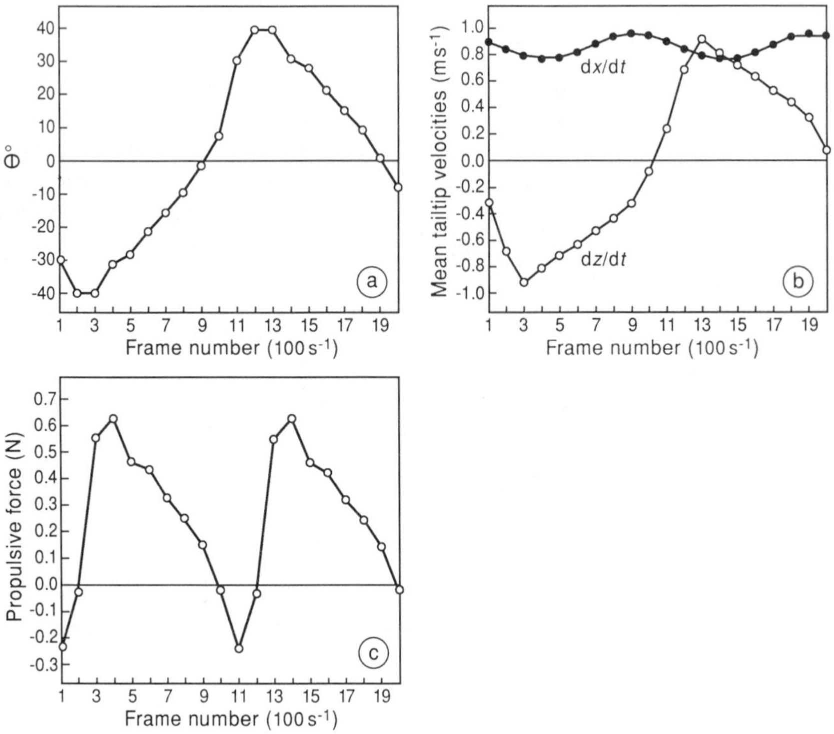

Fig. 1.8 Kinematic parameters of the end of the tail blade and instantaneous propulsive forces during one average tail stroke of a steady swimming bout at an average velocity of 0.86 m s^{-1} of a 0.3 m cod. See Table 1.2 for measured and calculated values.

exerted by the fish swimming at uniform speed. The mean propulsive force in the example of Fig. 1.8 is 0.26 N. The product of the average value of P and the mean velocity u is the average total energy output used to propel the fish (in W) involved. Our 0.30 m cod uses 0.22 W. The fish wastes kinetic energy into the vortex wake. The tail blade sheds $0.5mw^2$ (J m^{-1}) per unit length at velocity k, wasting energy at a mean rate of $0.5mw^2k$ (W); 0.08 W in the example. The propeller efficiency equals:

$$\eta_p = \frac{\overline{uP}}{\overline{uP} + \overline{(\frac{1}{2}mw^2k)}_{tail}} \tag{1.19}$$

The propeller efficiency of the steady swimming bouts of cod analysed in Fig. 1.8 is 0.73.

Table 1.2 Average values of swimming parameters of a 0.30 m cod (tail height s = 0.06 m and virtual mass m = 3.1 kg m^{-1} are treated as constants). See text for explanation of the calculations

Frame no.	dx/dt ($m\ s^{-1}$)	dz/dt ($m\ s^{-1}$)	Φ (*rad*)	Φ (°)	w ($m\ s^{-1}$)	k ($m\ s^{-1}$)	mwk*sin*Φ (*N*)	*0.5*mw^2*cos*Φ (*N*)	P_{av} (*N*)
1	0.90	-0.29	-0.53	-30	0.21	0.92	-0.30	0.06	-0.24
2	0.84	-0.69	-0.70	-40	0.01	1.09	-0.03	0.00	-0.03
3	0.79	-0.93	-0.69	-40	-0.21	1.20	0.50	0.05	0.55
4	0.77	-0.81	-0.54	-31	-0.30	1.07	0.52	0.12	0.64
5	0.78	-0.71	-0.49	-28	-0.25	1.02	0.38	0.09	0.47
6	0.83	-0.63	-0.37	-21	-0.29	1.00	0.32	0.12	0.44
7	0.88	-0.53	-0.27	-16	-0.27	0.99	0.22	0.11	0.33
8	0.94	-0.43	-0.16	-9	-0.27	1.00	0.14	0.11	0.25
9	0.96	-0.32	-0.02	-1	-0.30	0.97	0.02	0.14	0.16
10	0.94	-0.07	0.13	8	-0.20	0.93	-0.08	0.06	-0.02
11	0.90	0.29	0.53	30	-0.21	0.92	-0.30	0.06	-0.24
12	0.84	0.69	0.70	40	-0.01	1.09	-0.03	0.00	-0.03
13	0.79	0.93	0.69	40	0.21	1.20	0.50	0.05	0.55
14	0.77	0.81	0.54	31	0.30	1.07	0.52	0.12	0.64
15	0.78	0.71	0.49	28	0.25	1.02	0.38	0.09	0.47
16	0.83	0.63	0.37	21	0.29	1.00	0.32	0.12	0.44
17	0.88	0.53	0.27	16	0.27	0.99	0.22	0.11	0.33
18	0.94	0.43	0.16	9	0.27	1.00	0.14	0.11	0.25
19	0.96	0.32	0.02	1	0.30	0.97	0.02	0.14	0.16
20	0.94	0.07	-0.13	-8	0.20	0.93	-0.08	0.06	-0.02

1.6 SUMMARY AND CONCLUSIONS

Density is the most important physical property of water with regard to fish swimming. Viscosity is an important factor for small larvae, and plays a minor role in the hydrodynamics of adult fish, where it determines the flow patterns in the boundary layer close to the fish. The dimensionless *Re* number conveniently expresses the relative importance of these properties.

Archimedes' law, Newton's principles and Bernoulli's equation explain the main hydrostatic and dynamic forces and moments acting on a fish. A description of these physical principles is given and applied to fish.

The flow pattern around a fish is directly related to the forces involved in coasting or swimming. The equation for the drag on a rigid streamlined body can also be used for fish coasting with a straight body and adducted fins. Forces related to active swimming are much more complicated. Two approaches are used to explain how lateral undulations of body and tail, the dominant form of locomotion in fish, achieve propulsion by creating a backward directed jet of water. In the first approach, the action of the tail is compared to that of a beating wing and a semi-hypothetical three-dimensional pattern of vortices is presented. These vortices form a chain of rings through which the jet of water is threaded. A second, theoretical, approach describes how momentum is imparted to the water by the undulating body and how, according to Newton's third law, reactive forces from the water propel the fish.

This first chapter has concentrated on physical aspects, providing some feeling for what happens between fish and water during swimming. Let us now turn to the fish and study the form and functions of the anatomical details directly related to swimming.

Chapter two

The structure of the swimming apparatus: muscles

2.1 INTRODUCTION

Functional anatomists assume, as a working hypothesis, that organisms are adequately designed to perform the functions needed to live and reproduce. For many fish species, the survival value of the locomotion apparatus is so high that we may take for granted that it represents an efficient swimming device. We will study the anatomy of the relevant parts of the swimming machinery with this idea in mind.

Efficient swimming requires that:

1. most of the energy expended by the swimming muscles be turned into appropriate motions of the propulsive surfaces;
2. the generation of propulsive forces from the fish to the water leaves as little kinetic energy in the water as possible;
3. the transmission of forces from the water to the body of the fish be achieved with a minimum of energy loss through elastic deformations;
4. the resistance of the moving body be as small as possible.

These four requirements are directly related to the main components of the swimming apparatus of fish. If we compare a fish metaphorically with a submarine, these principal components may be called the engine, the propeller, the shaft and the hull. The engine generates forces that are transmitted by the shaft and the propeller into the water. Reactive forces from the water push along the same structures in opposite direction and order against the main body. The hull should be streamlined to avoid large drag penalties.

The majority of fish species swim by undulations of body and tail, powered by the lateral musculature; other species use paired or unpaired fins moved by intrinsic muscles. A fish body is divided by the median septum between head and tail into two lateral halves. The vertebral column with its spiny processes forms the main skeletal support of this septum. Rows of blocks of relatively short lateral muscle fibres are arranged in a complex way on the left and right side of the septum.

This chapter provides a description of the lateral musculature (the engine). Chapter 3 concentrates on the main body axis (the shaft) and the fins (the propellers). Chapter 4 deals with the shape and the surface structures of fish bodies and interprets aberrant forms and structures as special adaptations.

The main bulk of swimming muscles of fish consists of the lateral muscles and I will focus attention on the shape, the fibre structure, the fibre diversity and the innervation of these highly characteristic fish muscles.

2.2 LATERAL MUSCLES AND SEPTA

In fish, relatively short lateral muscle fibres are packed into blocks (myotomes) between sheets of collagen (myosepts). The myotomes are cone-shaped and stacked in a metamerical arrangement on both sides of the median septum. The colour of the muscle fibres may be red, white or intermediate in different locations in the myotomes. Red fibres are usually situated superficially under the skin. The deeper white fibres form the bulk of lateral muscles, and in some species intermediately coloured pink fibres are found between the two.

The purpose of the complex architecture of fish lateral muscles is not fully understood. We don't know the details of how active muscles bend the fish and power the swimming strokes. However, a large body of detailed knowledge about the structure and function of fish muscles has been amassed, and all we probably need is a new way of looking at the data to solve the problem. In this Chapter I wish to survey the structural aspects of our knowledge. In Chapter 7 physiological data will be added and in Chapter 8 an attempt will be made to synthesize a picture of the undulatory swimming machine.

The sections on gross morphology (pages 24–30) and on fibre structure and function (Section 2.3), give accounts of the present state of our knowledge. These accounts are based on a survey of the literature and on my own experience with dissection of myotomes and myosepts.

The shape of the myotomes

Man has been confronted with the challenging geometric complexity of the axial muscles ever since fish have been prepared as food. It is therefore surprising to find that the earliest accurate descriptions date from the first

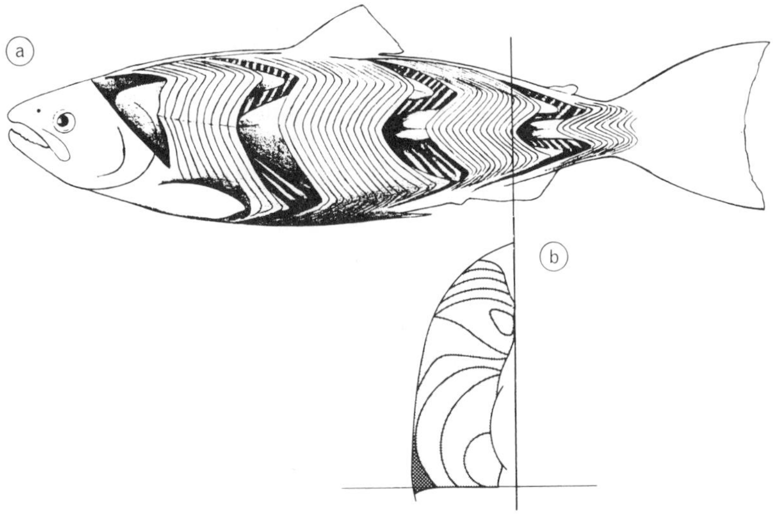

Fig. 2.1 (a) The shape of the myotomes on the left side of *Oncorhynchus tschawytscha*. Myotomes have been removed at four places to reveal the complex three-dimensional structure of the lateral muscle system. Redrawn and simplified after Greene and Greene (1913). (b) One quarter (top left) of a transverse section through the body of a young salmon, taken from the caudal region. The hatched area is occupied by red muscle fibres. The thick lines represent the myosepts between the stacked myotomes. Based on Shann (1914).

half of the 20th Century with important contributions by Greene and Greene (1913), Shann (1914), Kishinouye (1923) and Nishi (1938). The beautiful drawing of the myotomes of the king salmon, *Oncorhynchus tschawytscha*, of Greene and Greene became a textbook classic (Fig. 2.1(a)). Shann showed that in cross sections through the caudal region of teleost fish the muscles are arranged in four compartments: a dorsal and a ventral compartment on the left and right. The left and right halves and the dorsal and ventral moieties are mirror images of each other. A drawing of one quarter is therefore sufficient to illustrate the total cross section (Fig. 2.1(b)). Note the position of the red fibres on the outer side half-way down the body.

Lancelets (Amphioxiphormes) and lampreys (Petromyzonidae) lack a horizontal division (Harder, 1975a, b). In the other major groups, the dorsal and ventral moieties can easily be separated. There may or may not be a collagenous horizontal septum, sometimes it is absent in the caudal region only. In cod and mackerel for example the dorsal and ventral moieties are clearly separated, but a horizontal septum can hardly be distinguished. It may be prominent in other species.

Chevrel (1913) realized that the mathematical complexity of the shape of the muscles had to be related to the movements of fish and suggested that swimming movements would not be possible if the muscles were not segmented or if the myotomes were simple straight vertical blocks on both sides of the body axis. Experiments by van Raamsdonk *et al.* (1977) on the development of myotomes and myosepts in embryos of the zebra danio, supported this early view. He showed that embryos only started to move after the straight block-shaped somites became more complicated by folding backward.

Nursall (1956) used anatomical descriptions to make a comparison between the myotome patterns of fish-like chordates and tried to explain how these peculiar patterns of myotomes could cause bending of a fish during swimming movements. He emphasized the fact that attachment of each myotome to the vertical septum covers several intervertebral joints and that successive myotomes overlap. In his vision, shortening muscle fibres pull the myosepts and these pull obliquely to the long axis of the body on both sides across the overlapped vertebral joints, causing a local bending. The overlap of the myotomes allows the bending to progress smoothly along the body.

The tendon function of the myosepts was disputed by Willemse (1959), who provided arguments rather than proof to put forward the idea that collective shortening in longitudinal direction of muscle fibres on one side causes the fish to bend (he compared it with the way a bimetallic strip works). He believed that the myosepts only serve to connect the muscle fibres, end to end, between successive myotomes. Willemse (1966) pointed out that the oblique course of the myosepts and the complicated shapes of the cones are necessary to provide lateral space to allow the muscle fibres to increase their diameter during contraction.

A fellow-countryman of Willemse and mine, van der Stelt (1968), disagreed (in a thesis written in Dutch) with Willemse's dismissal of the tendon function of myosepts. He used elementary theories of bending to test various myotome models showing that myosepts were able to act as tendons in an efficient way. However, his analysis also indicated that the myosept patterns are more complex than they need to be. This finding is in contradiction with the starting point of this chapter, and my conclusion would therefore be that his method was not adequate to explain properly the complexity.

The fibre direction of the superficial red fibres is straight and longitudinal. Van der Stelt's anatomical studies of the lateral muscle fibres of the guppy, *Poecilia reticulata*, revealed that the fibre direction of the white fibres usually deviated from an orientation parallel to the body axis. This deviation was different in various parts of the myotome following the pattern shown in Fig. 2.2.

Independently and almost simultaneously, Alexander (1969) and Kashin and Smolyaninov (1969) were struck by the same feature. Alexander (1969)

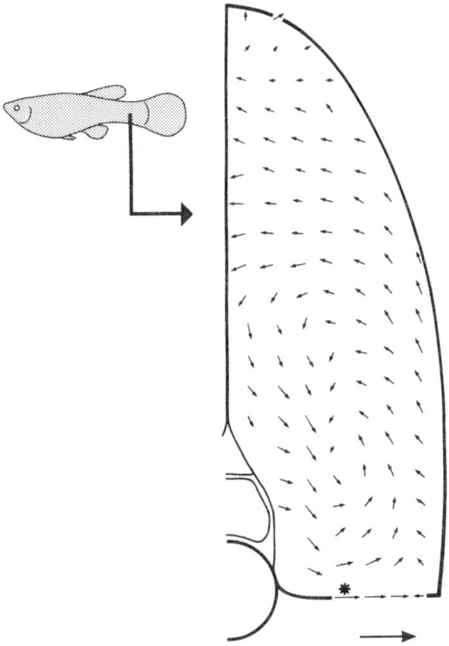

Fig. 2.2 The orientation of the muscle fibres in the lateral muscles of the tail in *Poecilia reticulata*, near the 25th vertebra. The arrows are projections on the transversal plane of vectors tangential to the muscle fibres. Redrawn with permission from van der Stelt (1968). Each vector is resolved into a component in the transversal plane, represented by the arrows, and a component parallel to the body axis. The length of this component has been chosen to be equal in all the places sampled and is represented on the same scale underneath. A muscle fibre parallel to the body axis is represented by an arrowhead; a fibre deviating at an angle of 45° from the direction of the body axis would be represented by an arrow as long as the arrow underneath. The muscle fibre represented by the largest arrow, indicated by an asterisk, deviates 25° from the direction parallel to the body axis.

discovered that the deviations were part of a complex three-dimensional pattern of white muscle fibre arrangement. Some fibres made angles as large as 40° with the long axis of the fish. (Later, Johnston and Salamonski, 1984, reported angles of over 80° for the Pacific blue marlin, Makaira nigricans.) Alexander managed to reduce the large variation to two main patterns. One exists in selachians and sturgeons as well as in salmonids and eels. It is restricted to the caudal peduncle of the other teleosts. The second pattern is generally found in the main part of the lateral muscles of the rest of the teleosts. In the selachian pattern the fibres are directed towards tendons in the top of the cones of the myotomes. The muscle fibres of teleosts in successive myotomes are arranged in helical trajectories if one follows a fibre straight through the myosept, continuing along the fibre directly opposite it on the other side of the myosept, and so on. Mathematical analyses by Alexander (1969) and van der Stelt (1968) suggest that, as a result of the oblique orientation, all the white muscle fibres in the myotomes are able to contract to a similar extent when the fish bends. For a given curvature of the body, the helically arranged white fibres contract with approximately one-quarter of the strain of the superficial red fibres (Rome *et al.*, 1988). This has important consequences for the way these muscles function, a matter discussed in Chapter 7.

Dissection of swimming muscles and myotomes

If one tries to skin a cod carefully, it becomes clear that the skin is firmly attached to the myosepts in some places, while it can easily be removed elsewhere. Fig. 2.3 provides a schematic picture of the tendon-inscriptions on the outside of the muscle mass. Towards the tail the angles between the lines made by the surfacing myosepts and the horizontal become smaller. The distance between adjacent myosepts decreases. On a cross section the myosepts show up as a series of lines giving away the fact that they have a complex spatial shape. It is very difficult to dissect myotomes and myosepts accurately in a freshly killed fish. It becomes more easy when a fish is boiled or, as I discovered by chance, even easier when a fish is smoked

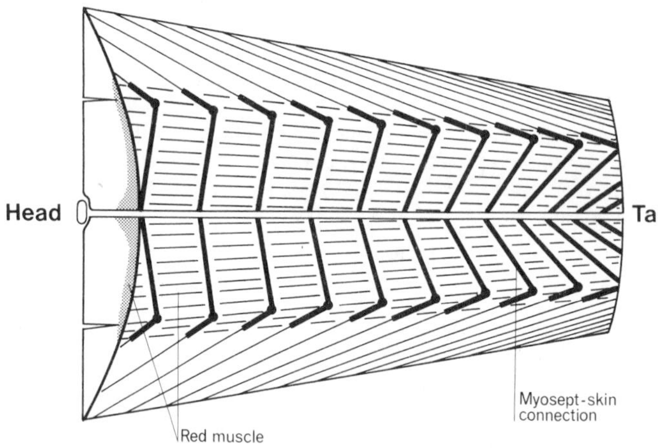

Fig. 2.3 Schematic drawing of the surface of a cod with the skin removed. The decrease of the distance between the myosepts towards the tail is slightly exaggerated to show how myosepts become more in line with the longitudinal axis towards the tail. Note how the area covered by red muscle coincides with the zone where the skin is firmly attached to the myosepts. Reproduced with permission from Wardle and Videler (1980).

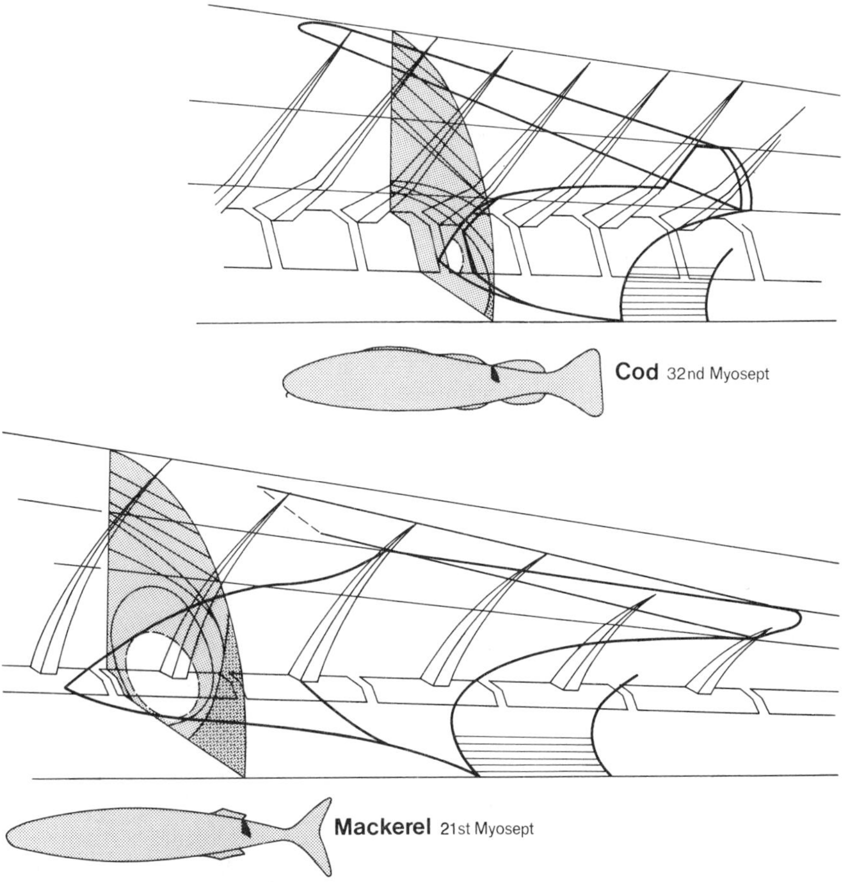

Fig. 2.4 Semi-schematic drawings in oblique projection of the 32nd dorsal myosept of cod and the 21st dorsal myosept of mackerel. The appearance of the myosepts on a transversal plane is also drawn for each species together with the area occupied by red muscle fibres represented by the more densely punctuated areas. The parallel horizontal lines indicate the width of the next myotome. Reproduced with permission from Videler (1985).

with the technique used for Arbroath smokies. The myosepts have disappeared and the muscle fibres form firm packages that can easily be separated. The myosepts of cod and mackerel in Fig. 2.4 are accurate reconstructions from smoked fish myotomes drawn in oblique projection. The regions occupied by red fibres are indicated on the transections. This drawing technique offers an accurate spatial relationship between the shape of the myosept and the vertebral column. It shows, for example, that the total length of the mackerel myosept spans more than five and that of cod slightly over four vertebrae. In both species the dorsal moiety consists of a smaller dorsal, and a large main part. A horizontal division can be found between these two parts. The septa of the dorsal part in cod are not continuous with the septa of the main part as is indicated in Fig. 2.4. The shaded transversal sections show the areas occupied by the red fibres as well as the myosepts indicated by lines that can be found in each of these transections. The myosepts of teleosts can be disappointingly weak, especially in the first half of the body. On the other hand, in the caudal peduncle of fast fish, such as tunas and mackerels, firm tendon-like structures are found. Myosepts of cod are stronger nearer to the skin in the region of attachment to the skin (Fig. 2.3).

Myosepts are in some species reinforced by intermuscular bones. These are the 'fish bones' encountered when eating fish. Shapes, locations and numbers of these bones vary in different species. For example, they are plentiful in cyprinids (on average 100 in carp, *Cyprinus carpio*, according to von Sengbusch and Meske, 1967), or may be completely absent.

2.3 DIFFERENT MUSCLE FIBRES

According to a reference by Bone (1966), it was Lorenzini who described in 1678 red and white muscle fibres for the first time in the elasmobranch *Torpedo*. In some fish this colour difference is very obvious but in many species the fibre arrangements are more confusing. The proportion of red and white fibres is related to life-style (Boddeke *et al.*, 1959). Active pelagic fish families have the highest proportion of red, whereas the lateral muscles of fish that spend most of their time lying on the bottom are virtually all white. The relative cross-sectional area occupied by red fibres changes along the fish from head to tail. In mackerel and herring, *Clupea harengus*, for example, it increases from between 5% and 10% just behind the head to 20% of the total cross-sectional area in the caudal peduncle. In saithe, *Pollachius virens*, the red muscle area increases from 1% near the head to about 8% of the total area in the tail region. However, this is not a general trend for all fish, because the percentage of cross-sectional area occupied by red fibres in swordfish, *Xiphias gladius*, and eel, *Anguilla anguilla*, reaches maximal values of about 7–8% at about 0.65 *L* from the nose. The largest amount of red muscle mass of the albacore (*Thunnus alalunga*) is found halfway down the body where it reaches values of 13% (He, 1986). Measurements of this ratio along the body from head to tail have only been made for a small number of species, and general rules or trends cannot be defined. The total volume occupied by red fibres is in the order of 10–14% of the total lateral muscle volume in mackerel and herring, approximately 5% in saithe and only 1% in cod (Videler, 1985; He, 1986). There is usually no septum separating the red and white fibres, but there are exceptions of which mackerel is an example.

A variety of techniques has been used to distinguish and test differences between fish muscle fibres. Morphological, histological, ultrastructural, histochemical, electrophysiological and biochemical studies provide different selection criteria, which could potentially make the number of fibre types infinitely large. There is more or less common agreement that slow red and fast white fibres are the main fibre types used in lateral undulations. Very often intermediately coloured pink fibres, with intermediate contraction speeds, can be recognized, usually situated between the red fibres on the outside and the more internal white ones.

Review articles by Bone (1978) and Johnston (1981) summarize the large body of knowledge on the structure and function of the three muscle fibres. I will first give an overview of the functional morphology of vertebrate striated muscle and subsequently discuss the different fish fibre types with emphasis on aspects important for our understanding of how undulatory swimming movements are powered.

The structure of vertebrate skeletal muscle fibres

Fish lateral muscle fibres show a number of structural and functional features common to cross-striated vertebrate skeletal muscle in general (reviews: Woledge *et al.*, 1985; Eckert *et al.*, 1988). The fibres are multinucleate and surrounded by a plasma membrane, the sarcolemma. The striated appearance is caused by the regular arrangements of thick and thin filaments in bundles called myofibrils. In the longitudinal direction, myofibrils consist of a sequence of identical units, the sarcomeres, interconnected by disc-like structures wrongly named Z-'lines' owing to their appearance as lines on longitudinal microscopic sections (Z originates from the German word *Zwischenscheibe*). The sarcomeres contain two kinds of filaments, which cause the contractile behaviour of the muscle. Thin filaments are attached to the Z-lines and are shorter than half the sarcomere length. Thick filaments run parallel to the thin ones and are shorter than the sarcomere length. The arrangements of the filaments are difficult to explain in words but are easily understood from a drawing as in Fig. 2.5. The regions where successive sarcomeres contain thin filaments are called I-bands in the cross-striated pattern because they are optically isotropic. The parts of the sarcomeres where the thick filaments as well as the overlapping thin filaments are found, form the A (anisotropic) bands on longitudinal sections. Sometimes, particularly in fish, the middle of the thick filaments is clearly visible as the M-line. The I-bands comprise, apart from the thin filaments, elastic connecting filaments, attached to the thick filaments and the Z-line. There are also cross connections between the longitudinal structures in the

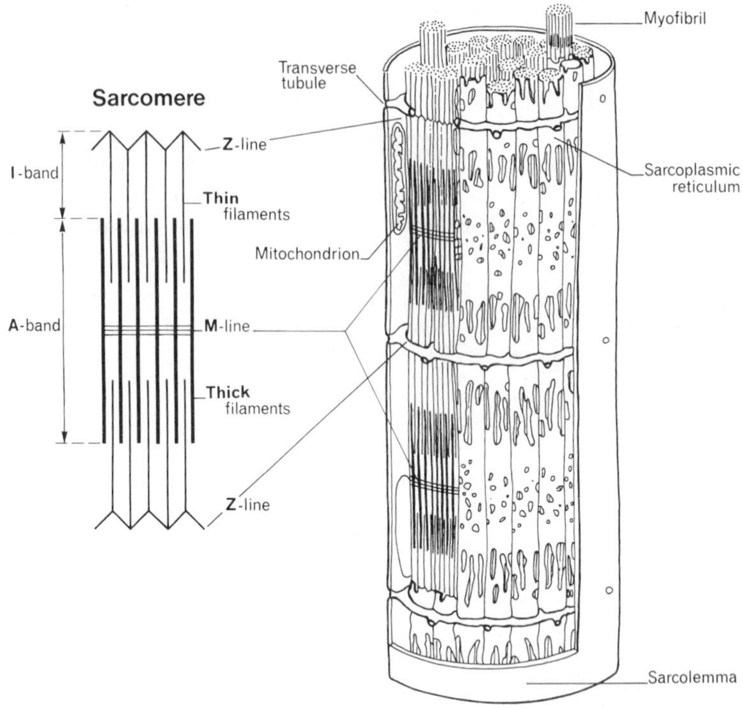

Fig. 2.5 Diagram showing the structure and nomenclature of a vertebrate muscle fibre. The fibre is surrounded by the sarcolemma, contains mitochondria and is densely packed with myofibrils. Each myofibril is in close contact with the sarcoplasmatic reticulum and consists of a series of sarcomeres. The cross-striated appearance of the fibres is caused by the isotropy of thin filaments (I-band) and the anisotropy of thick filaments (A-band) in the sarcomeres. The overlap between thick and thin filaments in a sarcomere increases when the muscle shortens during contraction, decreasing the distance between the Z-lines. A transverse tubule system, open to the world outside the fibre through holes in the sarcolemma, often coincides with the position of the Z-lines. The right-hand side of the diagram is redrawn from Woledge *et al.* (1985).

I-band. The longitudinal and transversal connections in the sarcomeres contain the complex proteins titin and nebulin. These cytoskeletal elements presumably enhance sarcomere structural stability during cycles of contraction and relaxation and provide passive elasticity against stretching (Wang and Wright, 1988; Akster *et al.*, 1989).

During contraction the thin filaments with actin molecules slide along the thick filaments where myosin is the active molecule. Cross bridges are cyclically formed between actin and myosin in an active muscle. These cross bridges generate forces dragging the thin filaments along the thick ones and shorten the sarcomere and/or create tension. In fact we still don't know exactly what happens there and how muscles contract. But we do know that a sudden occurrence of high calcium concentrations allows ATP to be converted into ADP. The breakdown of ATP provides the energy needed for the build-up of tension due to short-lived but forceful connections between actin and myosin. For immediate use the ATP concentrations are maintained by stores of phosphocreatine (PCr) splitting into phosphate, which is added to ADP to make ATP, and creatine. During prolonged activities red and white fibres use different pathways to restore ATP supplies. We will come to that in a moment.

First we need to find out how contractions are triggered. Holes in the sarcolemma are connected to a system of branched tubules, running predominantly transversely into the fibres, enclosing the packed myofibrils. These T-tubules are open to the world outside the fibre and appear at regular distances along the fibre, coinciding with the same part in each sarcomere. The T-tubules conduct electrical stimuli into the fibre. The myofibrils are also surrounded by a network of closed tubules, the sarcoplasmatic reticulum (SR). The lumen of these tubules is not continuous with the T-tubules, but there are connections between the two systems made by structures named terminal cisternae. The SR reduces the calcium concentration around the myofibrils which allows relaxation, but it can also suddenly release calcium, to activate the myofibrils, when triggered by signals from the T-tubules. The speed of activation depends therefore on the intensity of the contact between T-tubules and SR (Akster, 1981, 1985).

Mitochondria are present between the sarcolemma and the packed myofibrils. These make ATP from ADP and inorganic phosphate through the complex process of oxidative phosphorylation. This is an aerobic process, which uses the oxidation of fat, protein or, in some species, glycogen as energy source (van den Thillart, 1986). Red fibres contain many mitochondria and mainly use this probably rather slow process. White muscle fibres have few mitochondria. These predominantly use hydrolysis of stored PCr to maintain ATP concentrations and subsequently anaerobic pathways, where glycogen is turned into lactic acid. This may supply energy more rapidly, but has two important disadvantages. The net ATP production per mole glucose is only a fraction of that through the aerobic pathway and lactic acid has to be excreted and oxidized and/or has to be oxidized *in situ* before the muscle can be active again.

Fish axial muscle fibres

Differences between three fibre types in fish have been shown to exist even at the molecular level (review: Johnston and Altringham, 1991). Van Raamsdonk *et al.* (1980) and de Graaf *et al.* (1990a) used specific antibodies against

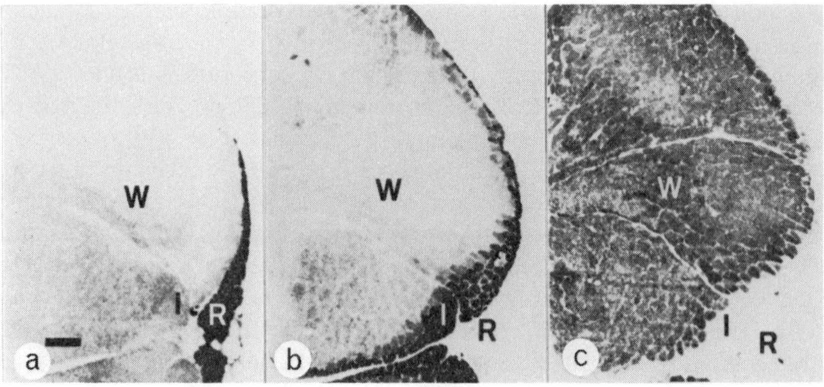

Fig. 2.6 Specific immuno-histochemical staining of red (R), intermediate (I) and white (W) muscle fibres in transverse sections of *Brachidanio rerio* myotomes. Antibodies against red, intermediate and white fibres are used to accentuate these fibres in panels (a), (b) and (c) respectively. Scale bar 250 μm. Photographs from De Graaf *et al.* (1990a) with kind permission.

three different types of myosin to demonstrate the position of red, white and pink fibres in the zebra danio myotomes (Fig. 2.6).

The diameter of the red fibres is only about 20–50% of that of the white fibres. Red fibres are commonly found on the outside, concentrated near the horizontal division between dorsal and ventral moieties. In some fast elasmobranchs and teleosts the red fibres are found more interiorly towards the vertebral column, an arrangement that serves to raise the temperature of the swimming muscles above ambient values (page 37).

Contractions of the red fibres are slow but they are virtually inexhaustible, owing to their aerobic metabolism. The red colour is directly related to the stamina of these muscles because it is caused by an extensive blood supply and by the abundant presence of myoglobin. The muscle fibres contain high concentrations of large mitochondria. The activity of oxidative enzymes in red fibres is higher than in white ones. For example, the red fibres of the zebra danio contain four times as much succinate dehydrogenase (SDH, an enzyme of the Krebs cycle) as the white ones (de Graaf *et al.*, 1990b). The red fibres are capable of fast (in the order of minutes) recovery from high lactic acid concentrations.

The bulk of fish muscles consists of fast white fibres. They have large diameters of up to 300 μm. The white colour is caused by the poor vascularization and by the lack of myoglobin. Each fibre contains a few small mitochondria and uses anaerobic metabolism as the energy source. The fibres are fast but become exhausted within a short period of time. They take up to 24 h to remove the lactic acid after an all-out burst of activity.

Intermediate pink fibres, if present, are situated as a distinct layer between the red and white fibres. Their metabolism is aerobic, showing generally intermediate characteristics between white and red.

The length of the filaments in the sarcomeres of axial muscles in the carp has been measured by Sosnicki *et al.* (1991). They found thick filaments with average lengths of 1.52 ± 0.009 μm and 1.50 ± 0.037 μm (means ± SD) and thin filaments of 0.96 ± 0.009 μm and 0.97 ± 0.023 μm in the sarcomeres of red and white muscles respectively. Van Leeuwen *et al.* (1990) measured similar values for red muscle of carp from different locations along the body. The figures for each filament type are so close that we can draw the conclusion that the filament lengths do not differ between red and white fibres: carp thick filaments are about 1.5 μm and thin filaments 1 μm long. In the perch, *Perca fluviatilis*, however, a subtype of red axial fibres has longer thin filaments than the white fibres and other red fibres. These aberrant red fibres form the innermost layer of red axial muscle and are therefore named 'deep red'. Longer thin filaments are not the only structural difference between the deep red and the other fibres. The position of the T-tubules is at the Z-lines in white and red fibres with relatively short thin filaments and at the junction between A- and I-bands, in the deep red fibres with long actin filaments (Fig. 2.7).

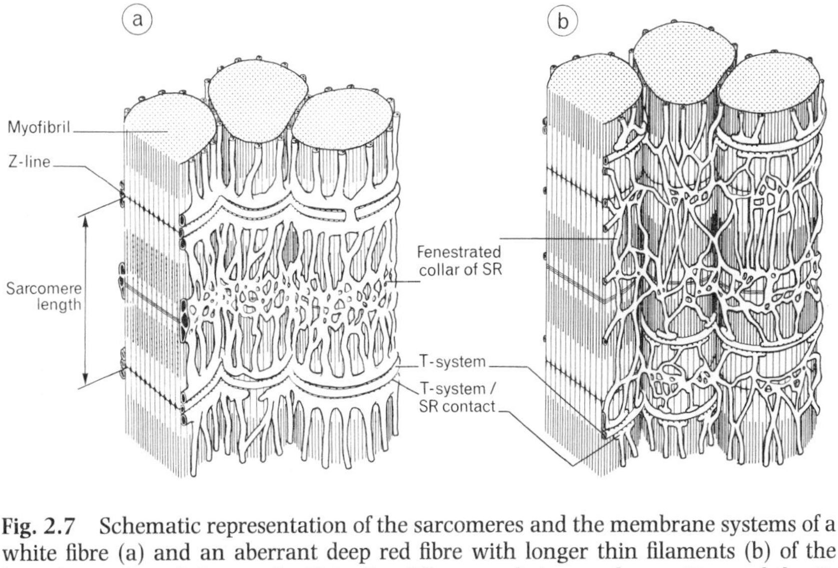

Fig. 2.7 Schematic representation of the sarcomeres and the membrane systems of a white fibre (a) and an aberrant deep red fibre with longer thin filaments (b) of the lateral muscles of the perch. Note the differences between the positions of the T-tubuli, the structure of the sarcoplasmatic reticuli (SR) and the intensity of the contacts between T-system and SR. Reproduced with permission from Akster (1981).

The extent of the contact between the T-tubules and the SR correlates with the fibre types and hence with different contraction velocities in carp and perch. The extent of the contact is small in red fibres, large in white and intermediate values are found for pink fibres (Akster, 1981, 1985; Akster *et al.*, 1985a, b). The same ultrastructural studies showed increasing thickness of Z-discs in red, pink and white fibres respectively.

Another special type of lateral muscle fibre is probably not involved in moving the body but is thought to play a postural role. Two species of bottom living sharks (the dogfish, *Scyliorhinus canicula*, and the closely related nursehound, *Scyliorhinus stellaris*) (Bone *et al.*, 1986) were shown to have a thin layer of thick fibres on the outside of the body. Kilarski and Kozlowska (1987) described similar fibres deep in the red zone of some freshwater teleosts. These tonic fibres are densely packed with large myofibrils. The diameter of the myofibrils is approximately four times that of the red, white and pink twitch fibres. The maximum contraction speed of these superficial fibres of the dogfish is only about one-third of the maximum speed of normal red fibres, and white fibres are maximally nine times as fast (Bone *et al.*, 1986).

The white muscle layer of salmonids is pink and consists of a mixture of large-diameter fast fibres and small fibres that are structurally different from red fibres. These are probably growth stages of large fibres. Harder (1975) mentions that the pink colour of salmon white muscle is caused by the deposition of astaxanthin, the colouring substance of the armour of gammarids. These crustaceans are supposedly the main food source of salmonids.

The enigmatic teleost thick filament arrangement

There is a peculiar and unexplained difference between muscles of teleosts and those of all other vertebrates. The arrangement of the thick myosin filaments in the A-band of teleosts shows on transection a simple lattice structure. X-ray diffraction reveals that all the heads of the myosin molecules on each of the thick filaments point in the same direction. The myosin heads on the thick filaments of all other vertebrates have one of two different orientations, 180° apart, which makes the repetition pattern more complex. The functional significance of this phenomenon has not yet been explained (Squire *et al.*, 1990).

Hot red muscle

The red muscles of tunas and mackerel-sharks (Lamnidae) are exceptional because these are positioned well inside the white muscle mass and connected to the blood vessels of retia mirabilia (Carey and Teal, 1966, 1969a, b). This arrangement acts as a countercurrent heat exchanger. The internal red muscles heat up venous blood while active and this warm blood loses its heat to cold arterial blood coming from the gills at ambient temperature. The result is that the arterial blood enters the muscles at a higher temperature than it would do without this system. The temperature of the body including the lateral muscles is kept at a fairly constant level, up to 10 °C over the ambient temperature, improving muscle performance drastically as we shall see in Chapter 7. Detailed reviews of different aspects of warm-bodied fish are provided by Stevens and Neill (1978) and Graham *et al.* (1983).

2.4 INNERVATION

Innervation of fish lateral muscle deviates substantially from the general pattern in vertebrate muscles. Vertebrates usually have either twitch or tonic fibres. The outer membranes of the tonic fibres are not able to conduct action potentials and are multiply innervated. There is only slow contraction after a series of stimuli. Twitch fibres conduct action potentials and produce a quick twitch contraction after one stimulus. These are focally innervated with one nerve terminal connected to the membrane near the end of the fibre.

Fish muscles fibres deviate from this simple dichotomy. Red fibres are slow but not tonic because they react to a single stimulus. Their outer membrane does not conduct the stimulus, but polyaxonal multiple innervation, with a high density of nerve terminals, makes the twitch response possible.

Innervation divides fast white twitch fibres of fish into two types. Until recently, the generally accepted idea was that older groups of teleosts and elasmobranchs had focally innervated white fibres (with either one or two motor axons per endplate at the fibre end) and that those of modern teleosts were all multiply polyneurally innervated along the fibre. In the bullrout, *Myoxocephalus scorpius*, for example, each fast fibre receives 8–20 endplates from 4 to 6 axons typically coming from the two adjacent spinal nerves (Altringham and Johnston, 1989). Raso (1991) has found the ancestral pattern in the catfish, *Ictalurus nebulosus* (brown bullhead), an advanced teleost. A closely related catfish, *Ictalurus punctatus* (channel catfish), displayed the multiple pattern. A functional explanation for these different patterns is lacking so far.

The spinal nerves of each myotome in the tail region of the zebra danio trifurcate into a dorsal, a ventral and a horizontal branch on both sides of the vertebral column. The dorsal and ventral branches run along the median septum and give off side branches to the dorsal and ventral moieties of the white muscle. The horizontal branch follows the horizontal septum to the pink and red muscle fibres (Fig. 2.8). The cell bodies of the motor axons running through these nerves lie in the spinal cord. Horseradish peroxidase

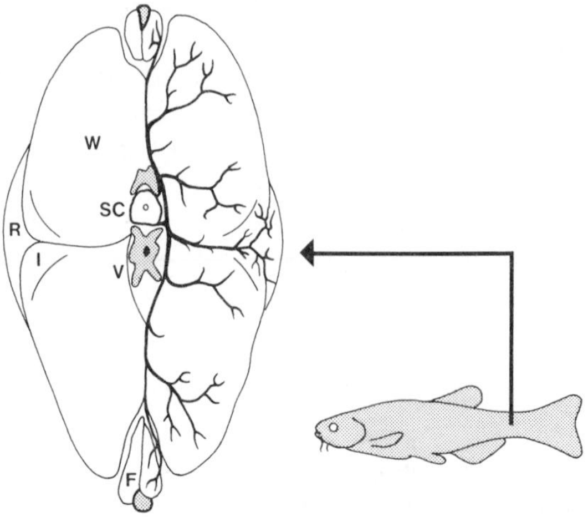

Fig. 2.8 Reconstruction of the nerves from the ventral root of the spinal cord, based on camera lucida drawings from silverstained series of sections through the tail of a zebrafish. SC, spinal cord; V, vertebra; F, fin muscle; R, I and W red, intermediate and white muscle. Reproduced with permission from De Graaf *et al.* (1990b).

injected into a muscle has the property to label axons from the muscle to the cell body. This retrograde tracing technique revealed that there are separate clusters of cells of motor neurones for red, intermediate and white muscles in the spinal cord. The clusters for red and intermediate are close together at some distance from the group of cells serving the white muscles (De Graaf *et al.*, 1990). Among the white motor neurones two types are distinguished: large primary motor neurones and smaller secondary ones. The white muscle fibres in the zebra danio are usually innervated by one primary motor neurone from the spinal nerve which has its origin in the same myotome, and receive some input from secondary motor neurones (Westerfield *et al.* 1986). We know little about the patterns of innervation that pass through these neurones, but the complexity of the white motor neurone system suggests that white fibres can participate in a wide range of locomotor activities. They are probably not only used during forceful activities at high speeds, but also gradually recruited to power cruising at increasing speeds. Physiological studies in Chapter 7 will provide more evidence for this hypothesis.

2.5 SUMMARY AND CONCLUSIONS

Myotomes are the laterally symmetric, metamerically arranged blocks of lateral muscle fibres in fish, subdivided in a dorsal and a ventral portion on each side. In the tail region the lateral halves and the dorsal and ventral moieties are mirror images of each other. The myotomes are usually double cone shaped. The fibre direction deviates slightly from the longitudinal direction. The angle of deviation depends on the position with respect to the vertebral column. The collagenous tissue of the myosepts separates and connects the myotomes. Myosepts may be strongly developed in places (even reinforced by fish bones in some species) and surprisingly weak elsewhere.

The colour of the muscle fibres in the myotomes is red, white or intermediate pink. Usually the bulk of the fibres is white; the red fibres form a thin layer on the outside just beneath the skin. Pink fibres may be found between the red and white ones. The colour and other structural differences between red, intermediate and white fibres are directly related to differences in function. The red fibres are slow but virtually inexhaustible and their metabolism is aerobic. The white fibres are fast, exhaust quickly and use anaerobic metabolic pathways. Pink fibres are intermediate in most aspects.

Fish red fibres are slow but, in contrast to other slow vertebrate fibres, react to a single stimulus because of the high density of nerve terminals on the fibres. White fibres are either focally or multiply innervated, in the zebra danio by one primary and some secondary motor neurones. Motor neurones from branching spinal nerves innervate the fibres in one to three adjacent myotomes.

At all structural levels, fish lateral muscles and nerves show unique features. We expect all these to be related to optimizing swimming, but in most cases the question remains how. Most fish appear to have two separate engines, a red one for slow swimming and a white engine for fast bursts. The white muscles occupy most of the fish body. Unlike terrestrial animals, fish can afford to carry this emergency system about, probably because it is virtually weightless and its bulk helps to streamline the body.

Chapter three

The structure of the swimming apparatus: body axis and fins

3.1 INTRODUCTION

A median septum divides fish behind the head into two lateral halves. It consists of a sheet of collagenous connective tissue reinforced by the axial skeleton. At the distal end the skeletal part of the septum is modified to accommodate the attachment of the tail fin. The axial skeleton includes the vertebral column and the skull.

The vertebral column must have two, almost mutually exclusive, swimming-related mechanical properties:

1. it should be flexible in lateral directions to allow undulatory swimming motions.
2. it must be incompressible in the longitudinal direction to receive the forward-directed propulsive force from the tail without large losses through elastic deformation.

Fins of fish are used for propulsion, braking and steering. Their structure should allow both a subtle interaction with the water and the exchange of large forces. Fin rays, widely different in shape, are the basic elements of the fins.

The basic design of the rays and of the main fin types will be described in detail. The structure of fin rays is a unique feature of fish, revealing an extreme adaptation to the interaction with water.

3.2 THE MEDIAN SEPTUM AND VERTEBRAL COLUMN

The number of vertebrae in the vertebral column is within certain limits characteristic for a given species or at least genus. For example cod have about 53, European eel, *Anguilla anguilla*, 114 and mackerel 31 vertebrae. The number may, however, vary considerably within a species: herring, for example, has between 54 and 58 vertebrae (Harder, 1975a, b). Among the Perciformes the number varies between 23 and 40, among Cypriniformes between 25 and 50. Balistiformes have only about 15 vertebrae (Ford, 1937).

Fish vertebrae are usually amphicoelous, which means that they are diabolo-shaped. The space inside two adjacent concave ends is filled with notochordal tissue. The vertebrae are interconnected by layers of connective tissue, drawing them together and enclosing the notochordal tissue. In fish no articular surfaces comparable to those of tetrapods exist between the vertebrae (Harder, 1975a, b).

Vertebrae fitted with different types of processes are distributed over the length of the column from head to tail. All vertebrae possess a neural arch and spine on the dorsal side. In the abdominal region, lateral projections (apophyses) are connected with the ribs enclosing the abdominal cavity. The first two or three vertebrae just behind the head are usually modified to permit a proper connection with the skull and for various other reasons not related to locomotion. The caudal region begins where the ribs are absent and a haemal arch and spine are present on the ventral side of the vertebrae (Hollister, 1936). The neural and haemal arches and spines provide approximate dorsoventral symmetry in the caudal region. The angles between the dorsal and ventral spines and the body axis change along the body, reaching different values for different species. It is important to note that these inclination angles are always directed backward and never approach 90°. The dorsal spines of the 15 anteriormost vertebrae of the rudd, *Scardinius erythrophthalmus*, for example, make angles between 55° and 65° with the body axis. From there rearward, the angle increases to values of 75° at vertebra 20 where the caudal region begins. It gradually decreases from there to values as low as 35° at vertebra 33. The last three vertebrae (34–36) in the caudal peduncle of this species are strongly adapted for the attachment of the tail fin. The angles between the ventral spines and the body axis in the caudal region are smaller than those of the dorsal ones just behind the abdominal cavity, but the values are similar across the rest of the caudal region. This trend is also found in other fish of similar shape. Van der Stelt (1968) shows comparable values for tench, *Tinca tinca*, and perch. In extremely elongated eel-like fish, the angles between the spines and the body axis show less variation. The angles of the dorsal spines of the eel-pout, *Zoarces viviparus*, are between 45° and 50° in the abdominal region and 55°

over most of the caudal region, gradually decreasing to 45° near the caudal end (van der Stelt, 1968).

A distinct dorsal elastic ligament is threaded between head and tail through the neural arches of all fish species (it is in fact common in vertebrates in general) and a smaller ventral elastic ligament through the haemal arches of some species (Symmons, 1979). These ligaments add marginally to the elasticity of the vertebral column which is mainly determined by the elastic properties of the connective tissue between the vertebrae.

The bony or cartilaginous skeleton supports the collagenous part of the median vertical septum. The collagenous fibres run between dorsal and ventral spines mainly in one direction. The angles between these fibres and the body axis are directed forwards and are as large as the backward-directed angles of the spines. This was found in the guppy, carp and pike, *Esox lucius*, by van der Stelt (1968).

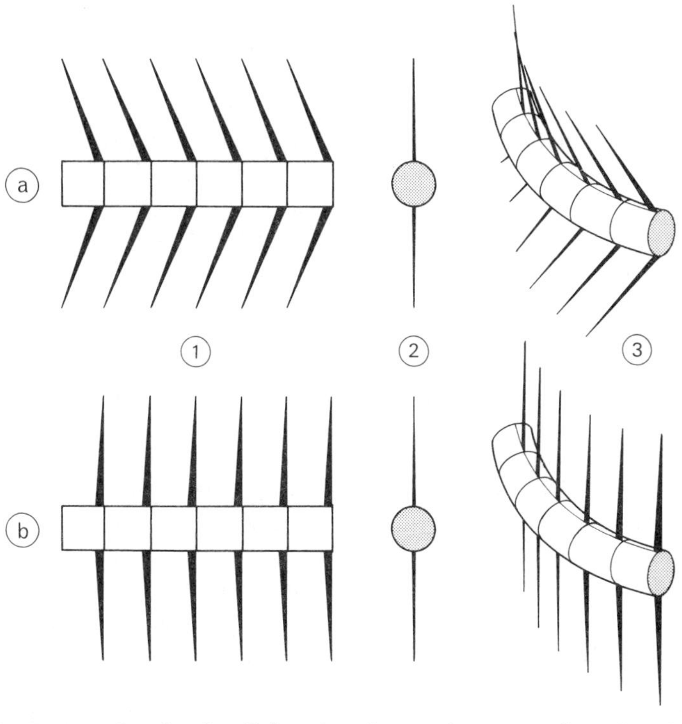

Fig. 3.1 Rearward-inclined stiff dorsal and ventral spines will tend to leave the median plane through the vertebral column during bending. This effect increases with increased angles of inclination; it does not occur if the spines are directed vertically. (a) Spines inclined, (b) spines not inclined. (1) lateral view, straight, (2) frontal view, straight, (3) drawing in perspective, curved.

Bending of the skeleton of the tail region of a fish will demonstrate that the obliquely inclined stiff spines leave the curved median plane through the vertebral column (Fig. 3.1).

The collagenous fibres in an intact vertical septum would have to lengthen to allow this to happen. Since collagen is hard to stretch, these fibres will tend to pull the spines back into the median plane. If that happens, the bone-and-collagen construction in the vertical septum acts as a spring, trying to restore the straight position of the fish. Such a spring could potentially save energy during rhythmic bending movements while swimming (Alexander, 1988). Although we have so far found three potential springs in the median septum (the elastic ligaments, the collagenous intervertebral connection and the collagen-stayed inclined spines), none of these is probably stiff enough to contribute substantially to saving energy used for swimming. Actual measurements of the elastic properties of the springs are needed to demonstrate whether this throw-away statement is right or wrong.

3.3 FIN RAYS

Fish fins are folds of skin, usually supported by skeletal elements. Fin rays are the skeleton of the distal part of the fin; they are connected to supporting skeletal elements inside the main body of the fish. Intrinsic fin muscles find their origin usually on the supporting skeleton and insert on the fin rays.

The fins of elasmobranchs are permanently extended and rather rigid compared with those of teleosts. The fin rays consist of rows of small pieces of cartilage called ceratotrichia.

Teleost fins are much more flexible than elasmobranch fins. They can be spread, closed and folded against the body. The design of the fin rays of teleosts probably represents the most elaborate and refined adaptation to efficient interaction with water that has ever evolved. There are two types of supporting elements: the lepidotrichia and actinotrichia. The lepidotrichia are supposedly rows of modified scales, growing internally (Goodrich, 1904). The very tip of teleost fins may be additionally stiffened by tiny horny rods, the actinotrichia, developed in the dermis of the skin. These are thought to be homologous with the ceratotrichia of elasmobranchs. Here, we are more interested in the analogy between the elasmobranch and teleost fin-supporting elements than in their ontogenetic and phylogenetic origins. The basic principle of how they function is probably similar, despite the large differences in movability. I shall start by explaining the form and function of elasmobranch fin rays and proceed from there to the more complex teleost fins. There are, however, as far as I know, no data on the exact form and function of ceratotrichia and so we shall have to restrict that discussion to teleost fin rays.

Structure and function of teleost fin rays

There are two kinds of teleost fin rays: spiny, stiff, unsegmented rays and flexible segmented ones. Spiny rays stiffen the fin and are commonly used for defence. The flexible rays play an important role in adjusting the stiffness and camber of the fins during locomotion.

The lepidotrichia are the important structures, the actinotrichia are usually extremely small.

Each lepidotrichium consists of two half rays or hemitrichia on each side of the fin (Fig. 3.2(a) and (b)). A hemitrichium has an unsegmented proximal part and numerous distal segments. The distal part can be either a single row of segments or it may be multiply bifurcated forming several parallel rows near the distal end. The two hemitrichia of one ray are exact

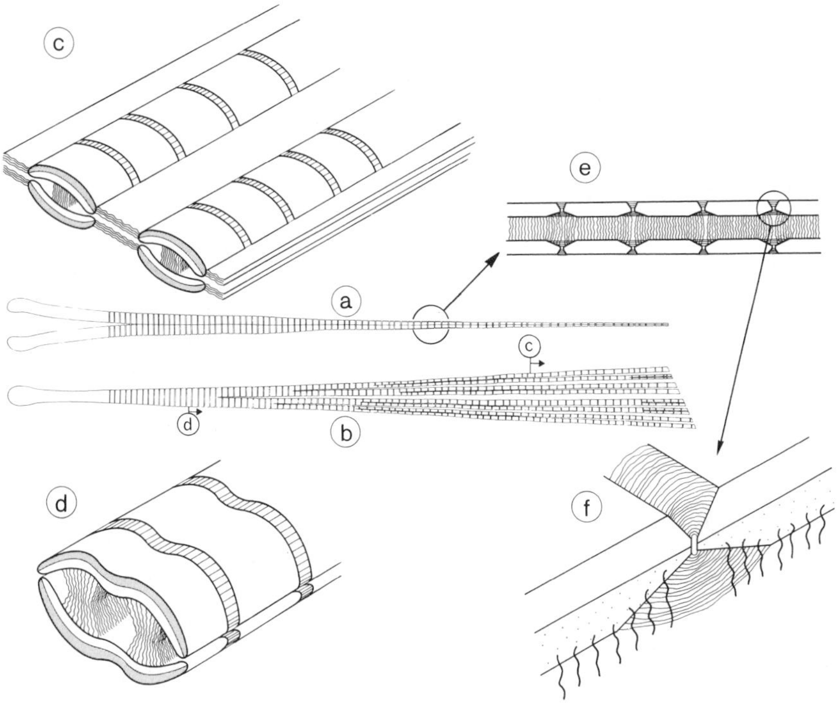

Fig. 3.2 Diagram of the skeleton and collagenous structures of a teleost fin ray. (a) Dorsal view; the hemitrichia are each other's mirror images. (b) Lateral view; note the irregular pattern of bifurcations. (c) and (d) Oblique views on transverse sections at positions indicated on (b). (e) Longitudinal section through the ray, at a position indicated on (a). (f) Oblique view of a longitudinal section through an intersegmental joint; an enlarged detail of (e). Reproduced with permission from Geerlink and Videler (1987).

mirror-images of each other. The joints between the segments of the opposing hemitrichia are situated exactly *vis-à-vis*, thus forming double joints (Haas, 1962; Arita, 1971; Lanzing, 1976).

The actinotrichia can be found as horny spicules supporting the edge of the fin membrane. Their proximal ends lie between the distalmost segments of the hemitrichia (Haas, 1962; Becerra *et al.*, 1983).

Structure of a lepidotrichium

Instead of summing up structural and functional details of teleost fin rays in general, I should like to proceed by concentrating on one particular fin ray by way of an example. Fig. 3.2(a) and (b) is a drawing of the skeleton, in respectively dorsal and lateral view, of the third dorsal ray (D3) taken from the tail fin of the tilapia, *Sarotherodon niloticus* (Geerlink and Videler, 1987). Each hemitrichium consists of 448 bony elements, making a total of 896 in the entire ray. In one row of segments from base to tip, up to 90 segments are linked by intersegmental joints. The longitudinal rows of segments are

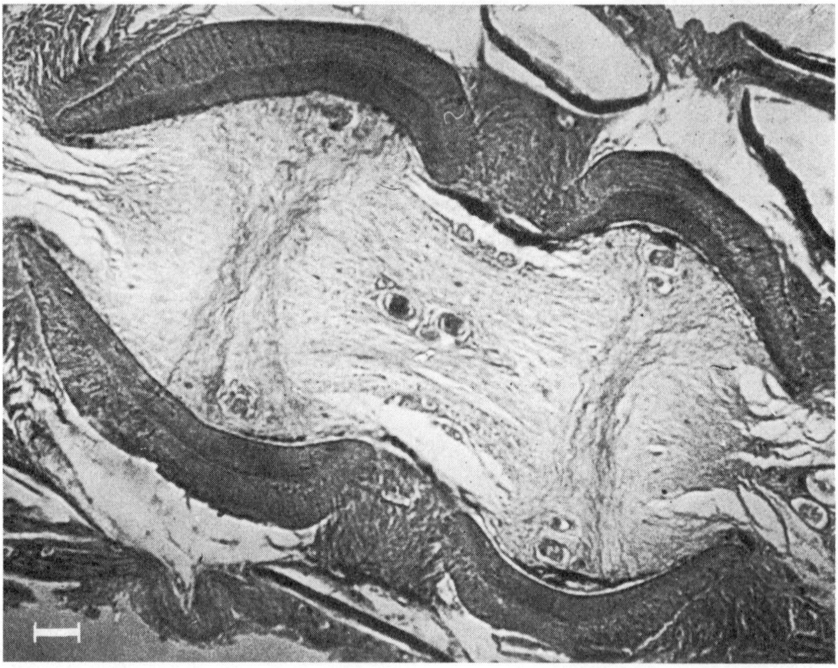

Fig. 3.3 Photograph of a transverse section of the fin ray schematically represented in Fig. 3.2(d). The scale bar is 0.15 mm. Reproduced with permission from Geerlink and Videler (1987).

connected by thick layers of collagenous fibres and covered by skin and small scales on both sides. Note in Fig. 3.2(b) that bifurcation does not occur in the same longitudinal position in each new branch. It does occur, however, in exactly the same way in the left and right hemitrichium, maintaining the mirror symmetry.

The length of the segments is approximately the same all along the ray. Fig. 3.2(a) shows how the ray tapers towards the end.

Cross sections (Fig. 3.2(c) and (d), Fig. 3.3) show the curved shape of the segments as well as the inner and outer bone layers with differently orientated collagen fibres. The outer bone layer is penetrated deeply by collagenous fibres connecting adjacent branches of the same or the neighbouring ray. From the inner bone layer, collagenous fibres run into the opposite segment, penetrating deeply. The serpentine, curly appearance of these transverse fibres is conspicuous. Longitudinal sections (Fig. 3.2(e), Fig. 3.4) illustrate how the serpentine fibres also penetrate into the intersegmental joints.

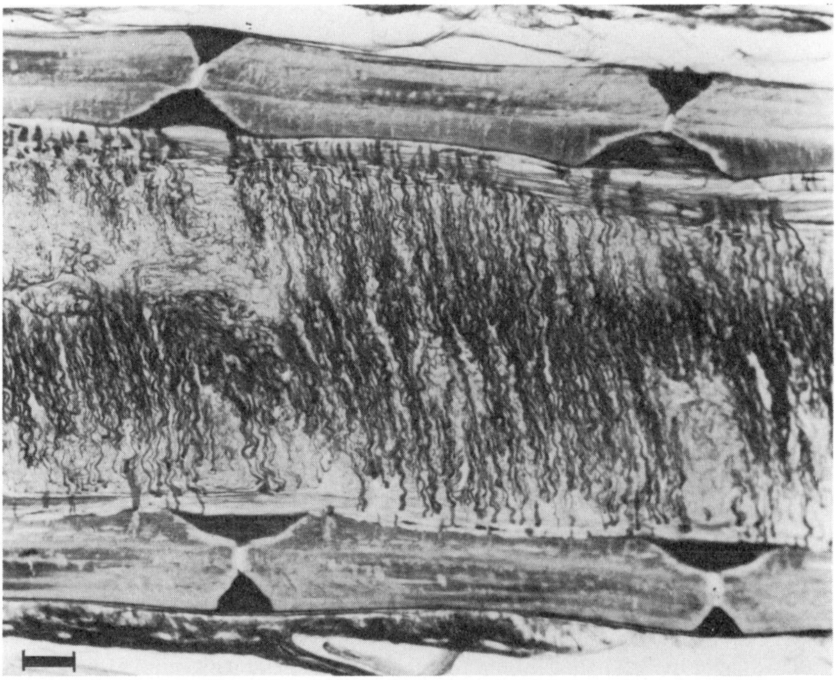

Fig. 3.4 Photograph of a longitudinal section through segments of the hemitrichia. Fig. 3.2(e) is the schematized representation of this section. The scale bar is 0.03 mm. Reproduced with permission from Geerlink and Videler (1987).

The ends of the segments on a longitudinal section appear to have the shape of an asymmetric blunt wedge (Fig. 3.2(f), Fig. 3.4). The tops of the wedges of adjacent segments abut, forming a bone-to-bone connection. On the outer side of the segments the converging angles of the wedges are smaller than on the inner side. The inner and outer dark triangles formed by the converging angles are filled with slightly waved collagenous fibres running parallel to the longitudinal axis of the ray.

Bending properties of the fin ray

Relative shifting of the hemitrichia causes bending of the ray (Fig. 3.5). The longitudinal displacement applied to the basis of one of the hemitrichia with respect to the other is proportional to the amount of bending. (The displacement is measured in mm or as a fraction of the fin ray length, and the degree of bending can be expressed as the surface area swept by the ray from its starting position, divided by the square of the fin ray length). The force

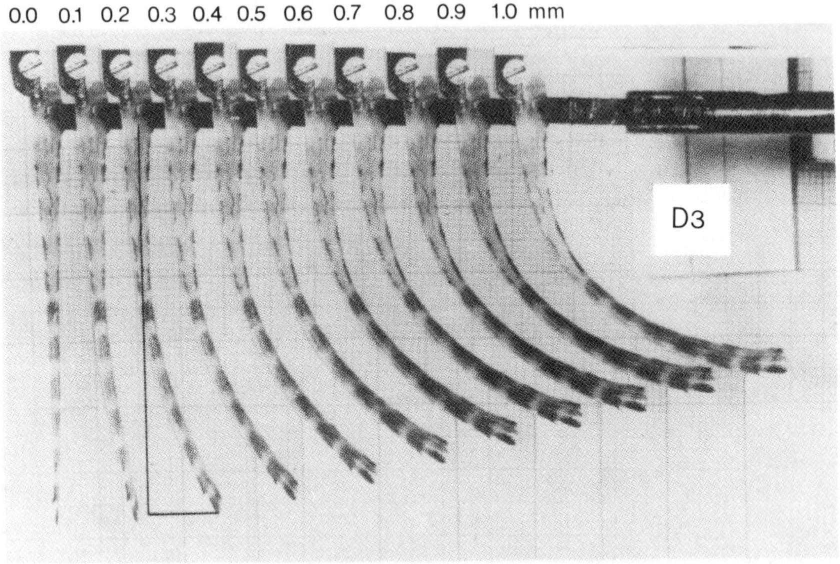

Fig. 3.5 Assembled photograph of the bending shown by a fin ray (D3 from the tail of tilapia) after the left hemitrichium was shifted at the base with respect to the right one over the distance indicated in each case. The clamp on the right-hand side was in fixed position and the left one was attached to a micromanipulator. The fin ray was taken from a freshly killed fish and kept underwater during the bending experiment. Reproduced with permission from Geerlink and Videler (1987).

needed to shift the hemitrichia increases exponentially with increased bending.

The distal end of the maximally curved fin ray in Fig. 3.5 points in a direction approximately 90° from the direction of the unsegmented basis of the ray. The 80–90 intersegmental joints would need to allow a rotation angle of about 1° each to reach this maximum. The slack in the collagen filling the wedges of the intersegmental joints allows for this angle.

How does longitudinal shift, applied at the basis, bend the distal part of the ray? McCutchen (1970) tried to answer this question without exact knowledge of the anatomy of the system. His explanation assumed that the distance between opposite hemisegments should not be able to decrease and that the tip segments were effectively joined. The curved shapes, on transection, of the segments (Fig. 3.2 (c) and (d)) show that his first assumption was correct, but the second one was only partly correct. Geerlink and

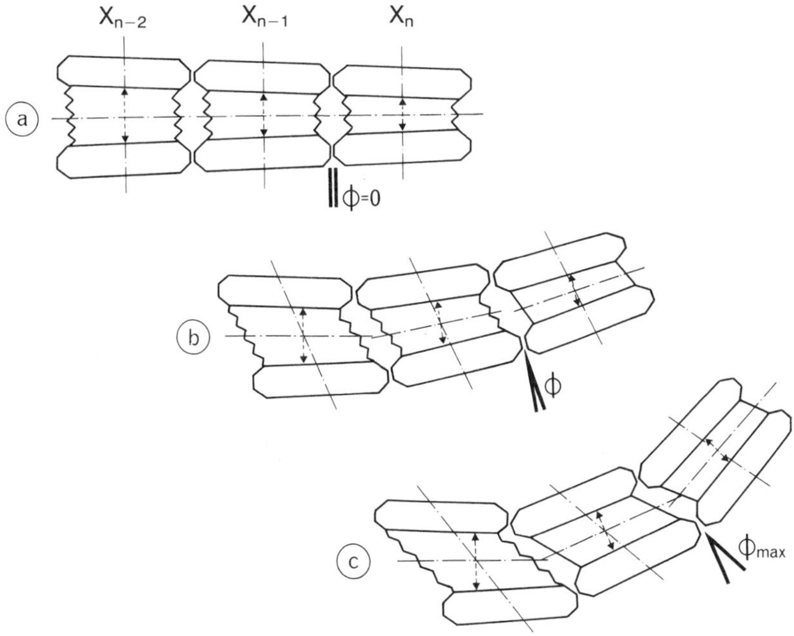

Fig. 3.6 Diagram of a longitudinal section through three juxtaposed segments of the hemitrichia of a fin ray. The zigzag lines connecting the segments represent the curly transverse collagenous fibres. ϕ is the angle of rotation between the segments after a longitudinal shift is applied between the hemitrichia in (b) and (c) to induce bending. (a) Neutral position, no shift; (b) transverse fibres stretched in segment X_n; (c) transverse fibres stretched in segments X_n and X_{n-1}. Reproduced with permission from Geerlink and Videler (1987).

Videler (1987) gave arguments favouring the idea that the curled serpentine fibres act as little springs. The intersegmental joints must also operate elastically according to their explanation. The serpentine fibres can be forcefully stretched to allow the hemisegments to shift with respect to one another. Once they are taut much greater forces are needed to stretch them. The distance between the hemisegments, and the length of the serpentine fibres, is smallest near the tip of the ray. Fig. 3.6 explains how distal bending by shift of the hemitrichia at the basis occurs. Three pairs of segments are shown at rest in (a). The differences in the dimensions are exaggerated to make the arguments clearer. The shift applied in (b) forcefully removed the slack from the serpentine collagenous fibres in the last segment and has pulled out those of the middle and first segment. The elastic forces in the system are also large enough to rotate the intersegmental joints over a small angle, which helps to accommodate the total relative shift. The shift has been increased in Fig. 3.6(c). The applied force is now used to pull the serpentine fibres taut in the third and middle segment, to stretch them submaximally in the first segment, and to rotate the intersegmental joints. The rotation in the last joint has reached its maximum value.

This whole complex system of bone and collagen provides an elegant balance of elastic forces distributed along the fin ray. Fish can control the stiffness and camber of each ray by shifting the hemitrichia at the basis. The joints and muscles responsible for this action will be described for each of the different fin types separately.

3.4 FINS

Having dealt with form and function of fin rays as the basic elements, we will now pay attention to the various fins in which these find an application. The skeletons and intrinsic muscles of the unpaired fins (tail, dorsal and anal fin) and the two paired appendages (the pectoral and pelvic fins) will be described and locomotion-related functions will be discussed. The emphasis will be on the fins of teleosts because detailed knowledge of form and function is hardly available for the fins of the other groups of fishes.

The caudal fin

Many species use the tail fin as the main propulsive and steering device. It is usually adapted to the needs determined by the type of swimming required to optimize survival and fitness in various niches. For slow-swimming fish in spatially complex environments (e.g. coral reefs or weedy rivers), steering capacity and manoeuvrability are more important properties than the ability to generate large propulsive forces. These fish need flexible and highly adjustable tails. At the other end of the range of variation we find the stiff and stout tails of fast pelagic swimmers. There are no obstacles in the open ocean where agility probably underlies high speed as a survival factor. These remarks are made to announce large variations in form and function of fish tails. There are a number of general trends in the form–function relations of caudal fins and I shall highlight these instead of aiming at covering most of the variation.

External shape and aspect ratio

The shapes of the outlines of side views of fish tails are either dorsoventrally symmetric or asymmetric (Fig. 3.7). The symmetric tails are called homocercal and the asymmetric ones heterocercal. Heterocercal tails are either hypocercal, if the lower half is larger than the upper one, or epicercal, in cases where the dorsal fin portion is extended. Sharks, with the exception of some of the fast pelagic ones, and sturgeons have epicercal tails. The thresher shark, *Alopias vulpinus*, is the most extreme example. Norman (1960) describes how the thresher uses this tail to feed on herring, pilchards or mackerel: 'It swims round and round a shoal of these fishes, thrashing the water with its long tail and thus driving the prospective victims into a compact mass, when they form an easy prey.' He does not provide hard evidence for this story but: *se non è vero, è ben trovato.*

During swimming the epicercal tail generates thrust in the horizontal swimming direction and also a vertical force directed upward. Experimental evidence for the vertical component in the thrust generated by epicercal shark tails was provided by Alexander (1965). Just how large the vertical force is depends on the relative stiffness of the different lobes of the fin (Simons, 1970). This generally accepted view seems logical because the upper lobe contains the extension of the vertebral column and will generally be stiffer than the lower lobes and lead in the direction of lateral motion. Thomson (1976), however, challenged this view and pointed out that resolution of the precise effect of the epicercal tail awaits investigations of the flow pattern around a swimming shark or sturgeon.

The hypocercal tail of flying fish, *Cheilopogon heterurus*, (Fig. 3.7) has a function during take-off from the surface. Hertel (1966) shows how the enlarged lower lobe of the caudal fin is still in the water and continues to beat after the body of the fish has already become airborne. Aleyev (1969) mentions the same phenomenon, quoting an article in Russian by Shuleikin (1928).

Teleosts predominantly have homocercal tails. These are dorsoventrally symmetric from the outside, but skeleton and muscles are often asymmetric. Thrust from these tails will be directed horizontally if the lateral movements

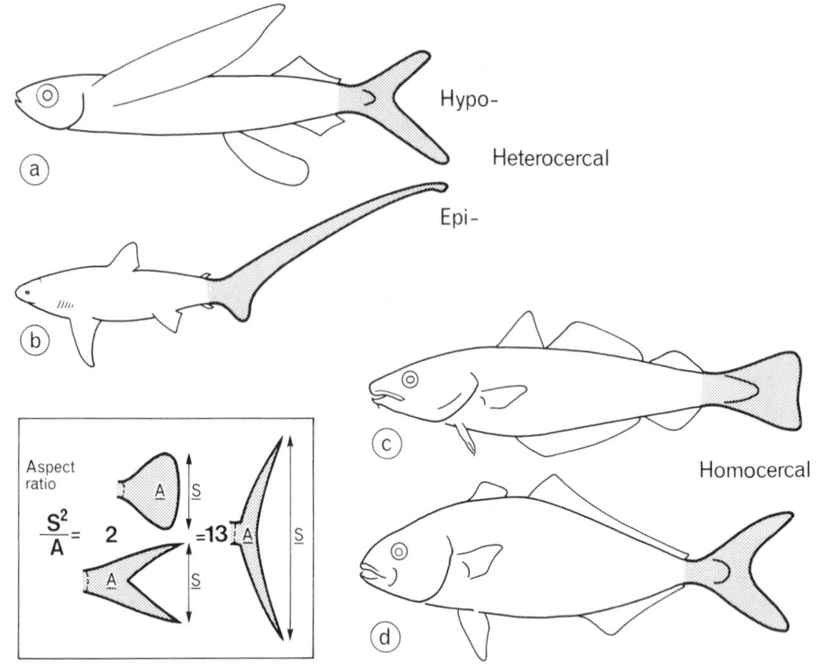

Fig. 3.7 Four examples of fish with different tail shapes. Two species with heterocercal tails are (a) the Atlantic flying fish, *Cheilopogon heterurus*, with an extended lower lobe (a hypocercal tail), and (b) the thresher, *Alopias vulpinus*, in which the upper lobe equals the body length (an epicercal tail). Two examples representing the variation among fish with homocercal tails are (c) the whiting, *Merlangius merlangus*, which has a low aspect ratio (AR), and (d) the amberjack, *Seriola dumerili*, which represents the fast swimmers with high AR tails. The inset explains the term aspect ratio (AR): the two tails at left have different shapes but the same low AR and that at right is a high AR tail. (S = tail height; A = surface area of the tail blade.)

are dorsoventrally symmetric. Using the asymmetric musculature, homocercal tails can make dorsoventrally asymmetric movements. Mackerel, for example, were observed to imitate epicercal movements, with the upper part of the fin leading the lateral tail sweep during slow swimming (He and Wardle,1986). I will come back to the reason for this behaviour when discussing form and function of the pectoral fins (page 64).

There is an enormous variety in shapes, ranging from almost circular to narrow, crescent-shaped tails. As a general rule, fast long-distance swimmers possess extremely lunate tails, with large spans and small chords. Fish that are highly manoeuvrable at short range have tails with a large surface area relative to the height of the fin. This variation can be conveniently expressed as the aspect ratio (AR), which is the height (or span) squared divided by the surface area (Fig. 3.7 inset). Fast swimmers have tails with the highest ARs. The amount of water affected is proportional to the span of the fin. The advantage of the large span of high AR tails is similar to that of the long narrow wings of modern glider airplanes. The narrow chords and well-designed cross-sectional profiles generate large pressure differences between the two surfaces of the hydrofoil in each case. This pressure difference causes the formation of the vortices at the tail tips. The vortices at the tips of long wings and high tails are small compared with the total length over which the pressure difference is maintained (page 13 and Fig. 1.4 (b)). That is the reason why the induced drag, caused by the energy dissipated in the tip vortices, of these tails is relatively small compared with that of tails with a smaller AR. In other words, high AR tails provide large thrust and low drag. The AR is not the only special feature of the tails of fast fish. The crescent shape, with backward-curving leading edges, is also typical. Comparisons of the hydrodynamic performance of tail shapes with the same AR, but with and without backward-curving leading edges, showed that a crescent shape reduced the induced drag by 8.8% (van Dam, 1987).

Skeleton, muscles and tendons of the caudal peduncle

The end of the vertebral column is usually adapted to accommodate the attachment of the tail fin. In some fish, in cod for example, the deviations from the general pattern of vertebrae bearing neural and haemal arches and spines are rather minute. Photographs of a clarified alizarine preparation, where the position of the skeletal parts is not in any way disturbed, show the adaptations in detail (Fig. 3.8). Length and diameter of the vertebrae gradually decrease from about 0.023 L near the head to 0.007 L at the end of the column. The vertebral column ends straight and the whole structure shows a strong dorsoventral symmetry. The fin rays are connected to a large number of modified neural and haemal spines. The last vertebra bears a triangular bony plate (the hypural bone), to which the four central fin rays are connected. It is in direct line with the vertebral column.

In many other groups the vertebral column ends as a large triangular plate. Tilapia provides a typical example in Fig. 3.9. Usually both neural and haemal spines are enlarged to form this plate. Haemal and neural spines belonging to vertebrae that have disappeared during phylogeny are called hypurals and epurals respectively. The hypurals usually play the most important role in the formation of the triangular plate, even in the upper half of the tail because the end of the vertebral column tends to be turned upwards. The upward-turning part may consist of a few fused rudimentary vertebrae, or of a styliform piece of notochord in some cases flanked or surrounded by bony elements. It is quite obvious in, for example, pikes,

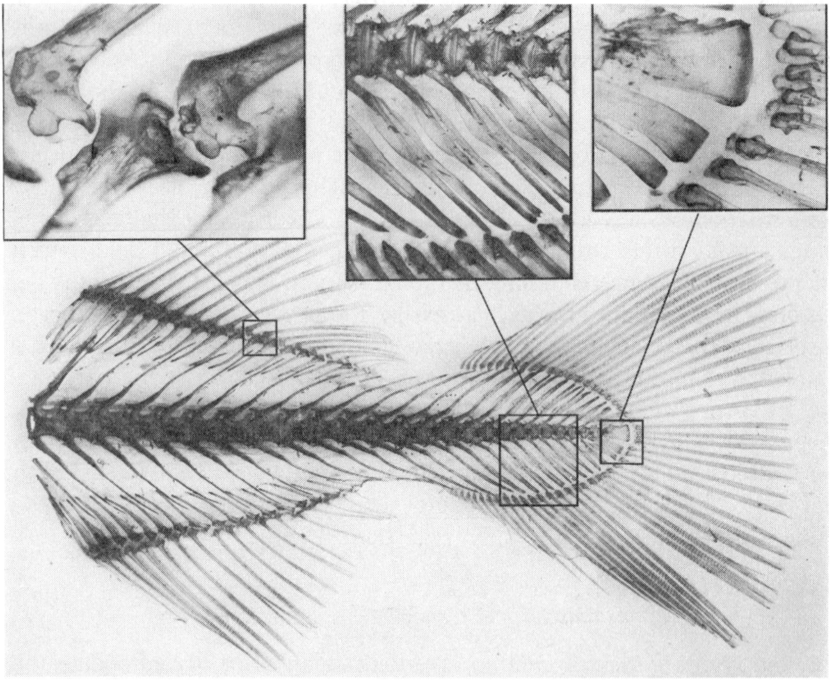

Fig. 3.8 Photograph of a clarified alizarine preparation of the tail of cod.

sticklebacks and salmonids, and moderately so in perciforms. The nomenclature of these elements (i.e. stegural, urostyle or pseudurostyle) is rather complex and is based on their precise shape and position and on the assumed phylogenetic origin (see Videler, 1975, for an overview of the names given to skeletal parts in the caudal peduncle). The last vertebra of perciforms bears lateral flanges projecting caudally, the hypurapophyses (Fig. 3.9). These important structures offer an origin for muscles away from the plane of the vertical septum (Nursall, 1963).

The musculature of the caudal peduncle usually consists of modified last myotomes, carinal muscles and a variety of intrinsic muscles. The extremely narrow caudal peduncles of fast pelagic fish are characterized by a lack of muscles and by long tendons from the last myotomes inserting on the fin rays. To illustrate the variety of musculature in the peduncle of a highly manoeuvrable teleost, I use the results of my dissection of the tail of tilapia drawn in Fig. 3.10. Dorsal and ventral marginal muscles can spread the rays, and series of interradial muscles can bring them closer together. The function of the supra- and infracarinal muscles is not clear. The origin of the hypochordal longitudinal muscles (HLM) extends from the last vertebra and

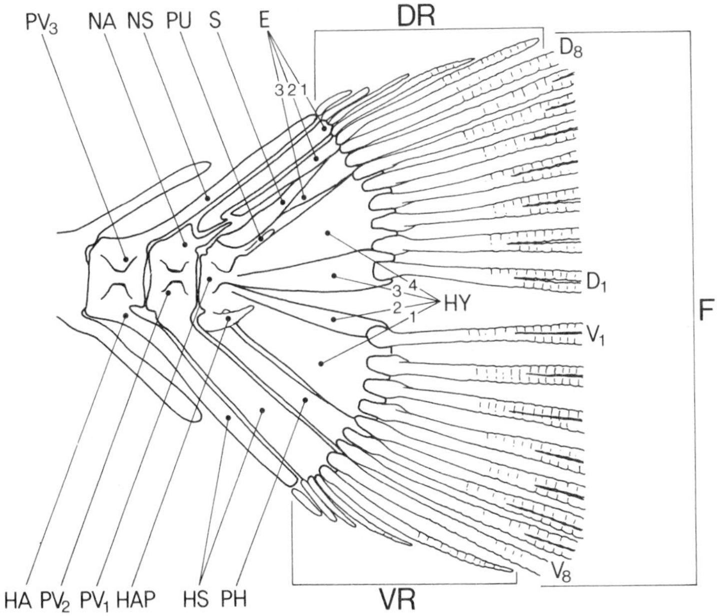

Fig. 3.9 Drawing of the caudal skeleton of tilapia, *Sarotherodon niloticus*. Abbreviations: DR, dorsal raylets (unbranched fin rays) (numbered D9–D13); E, epurals; F, fin rays (D1–D8 and V1–V8); HA, haemal arch; HAP, hypurapophysis; HS, haemal spines; HY, hypurals; NA, neural arch; NS, neural spine; PH, parhypural; PU, pseudurostyl; PV, preural vertebra; S, stegural; VR, ventral raylets (V9–V13). Reproduced with permission from Videler (1975).

central hypural plate outward onto the lateral flanges of the hypurapophyses. The HLM is the only muscle clearly disturbing the dorsoventral symmetry in the caudal peduncle of perciform fishes. Its fibres run at an angle of about 75° with the body-axis dorsocaudally inserting with four long tendons on the heads of fin rays D5–D8. The last myotomes of the lateral muscles insert on the fin ray heads of the 14 centralmost fin rays. A triangular sheet of collagen connects the lateral muscle with eight fin rays in the centre, leaving a gap in the dorsal part for the tendons of the HLM to emerge. Underneath the last myotome, a variety of dorsal and ventral muscles can be found. Usually there are two major bundles inserting on the heads of the 16 fin rays in the centre of the fin. Tilapia has a dorsal and a ventral chiastic bundle overlaying the major ones and inserting on two or three fin rays in the very centre of the fin. The variety of muscles gives away the versatility of the tilapia tail fin movements. Precise knowledge of separate functions of the individual bundles is, however, not easy to assess. The general picture emerging is that muscular force is transferred onto the

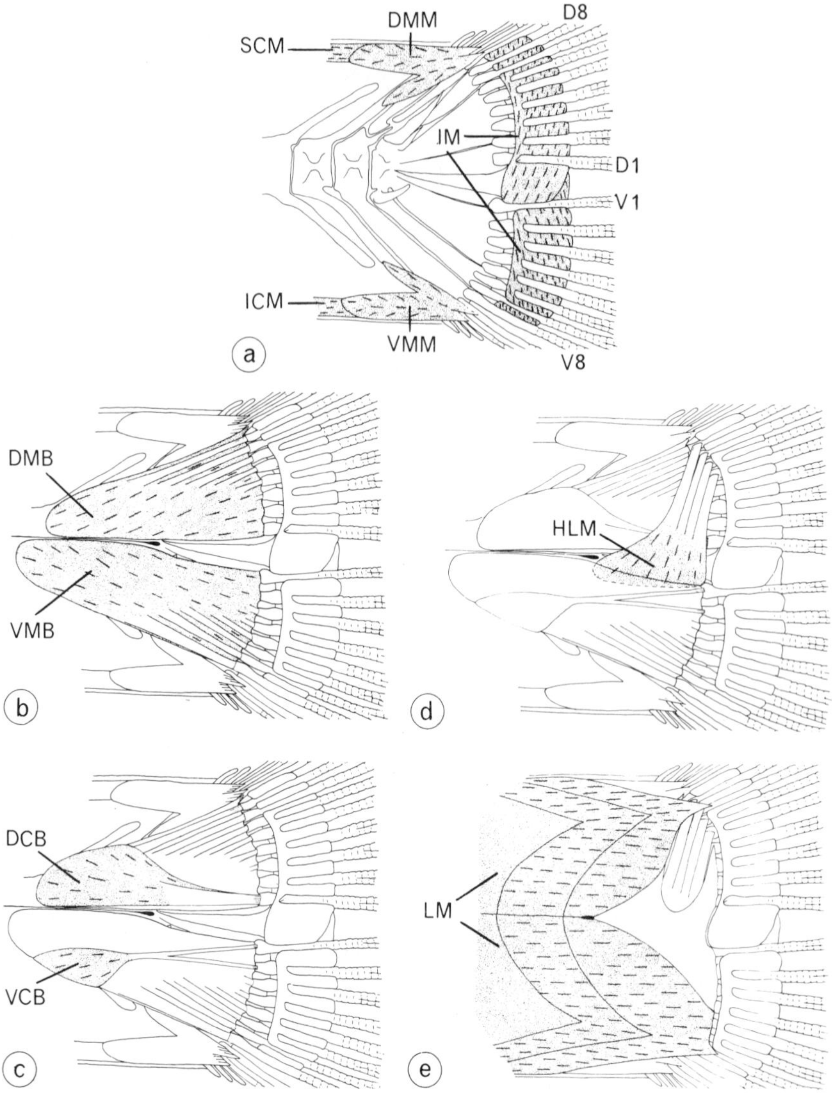

Fig. 3.10 The muscles of the tail of tilapia, *Sarotherodon niloticus.* (a) Marginal muscles; (b) and (c) profundal muscles; (d) and (e) superficial muscles. Abbreviations: DCB, dorsal chiastic bundle; DMB, dorsal major bundle; DMM, dorsal marginal muscle; HLM, hypochordal longitudinal muscle; ICM, infracarinal muscle; IM, interradial muscle; LM, lateral muscle; SCM, supracarinal muscle; VCB, ventral chiastic bundle; VMB, ventral major bundle; VMM, ventral marginal bundle. Reproduced with permission from Videler (1975).

fin ray heads. The angle of insertion and the exact position on the fin ray head of the insertion determine the contribution of the activity of each muscle to the overall movements caused by the muscles in concert.

Electromyography of the tail muscles during swimming of tilapia (Videler, 1975) reveals that intrinsic tail muscles are active up to 65% of the time during strokes to the ipsilateral side of the muscles. There is also considerable activity on both sides of the peduncle when the tail changes its direction of movement. The muscles on the contralateral side with respect to the direction of a stroke are active up to 12% of the stroke time. The intrinsic caudal fin muscles exert forces on the fin ray heads. We need to know more details of the connection between fin rays and peduncle to be able to appreciate what the consequences of the activities of the muscles are.

The fin–body connection

Horizontal sections through this region (cod as an example in Fig. 3.11) allow interpretation of the connection between axial skeleton and fin rays.

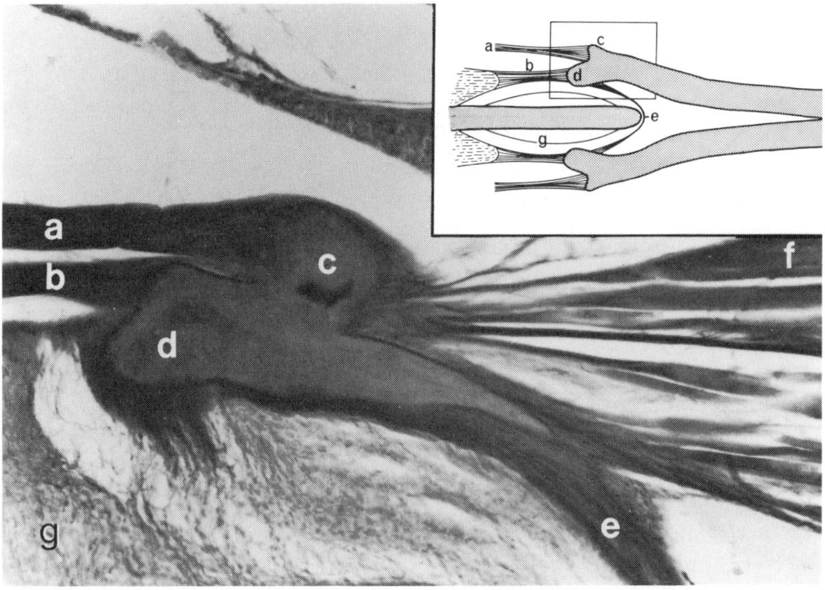

Fig. 3.11 Insertion on a caudal fin ray head of cod shown in a horizontal (frontal) section. a, Insertion of the skin; b, insertion of last myotomes and intrinsic muscles; c, lateral process on fin ray head; d, fin ray head; e, capital ligament; f, interradial muscles; g, cushion of connective tissue. The inset provides a diagram of the connection between the hypural plate and the fin ray. The body of the fish is to the left and the tail to the right. Reproduced with permission from Videler (1981).

The inset in Fig. 3.11 shows how a ligament (the capital ligament, Videler, 1975) between the heads of the hemitrichia locks the ray onto the plate.

Let us first discuss how this construction can transfer propulsive forces from the tail blade to the vertebral column. Thrust force, generated by the tail, pushes the heads of the hemitrichia forward. The capital ligament prevents forward shifting of the fin ray heads and pushes on to the hypural plate. The heads of the hemitrichia will be pressed inwards against cushions of connective tissue on both sides of the hypural plate. As long as thrust is being generated, the connection between ray and plate remains rather rigid.

The same construction is used to transfer muscular forces from the main body to the tail blade. The lateral oscillation of the body moves from head to tail and reaches the caudal peduncle where the intrinsic muscles and skin transfer the forces to the fin by pulling on one side of the fin ray bases.

When no thrust is generated, the tension on the system is released and the connection between fin ray and plate is loose and can easily be adjusted. The various possibilities for adjustment are outlined in Fig. 3.12((a) to (e)) (Videler, 1977). Forces on the joint, when the fin pushes the fish, are depicted in Fig. 3.12(b). Fig. 3.12((c) to (e)) summarizes the three possibilities for adjustment.

1. The angle between the longitudinal axis of the unsegmented part of the fin ray and the main axis of the body can be altered by the fish (Fig. 3.12(c)).
2. The fish can spread or close the tail fin as a fan. The axis of rotation for this movement is indicated in Fig. 3.12(d).
3. The longitudinal shift of the two hemitrichia (Fig. 3.12(e)) changes the stiffness and camber of the fin ray as discussed on page 48.

The last myotomes end tendinous and insert on the fin ray heads inside the strong attachment of the collagenous part of the skin (Fig. 3.11).

The connection between fin rays and vertebral column fulfils two largely incompatible functions. It must be highly adjustable if manoeuvrability is required, and rigid in case the transfer of large forces is the most important function. It is therefore not surprising to find large differences when comparing the joint of a highly manoeuvrable demersal fish such as the cod with that of the fast steady pelagic swimmers mackerel and tunny. In Fig. 3.13 schematic horizontal sections through the mechanically important structures are compared. The fin ray heads of mackerel partly overlap the ural plate, and collagen strongly connects the fin with the plate. The capital ligament is also very prominent and capable of transferring large forces. It has lost this function in the tunny, where the fin ray heads completely overlap the ural plate and abut directly against the last vertebral body. Propulsive forces are transmitted along this bone-to-bone connection, and hence without losses through elasticity, from the tail onto the vertebral column. The capital ligament prevents the fin ray halves from splitting

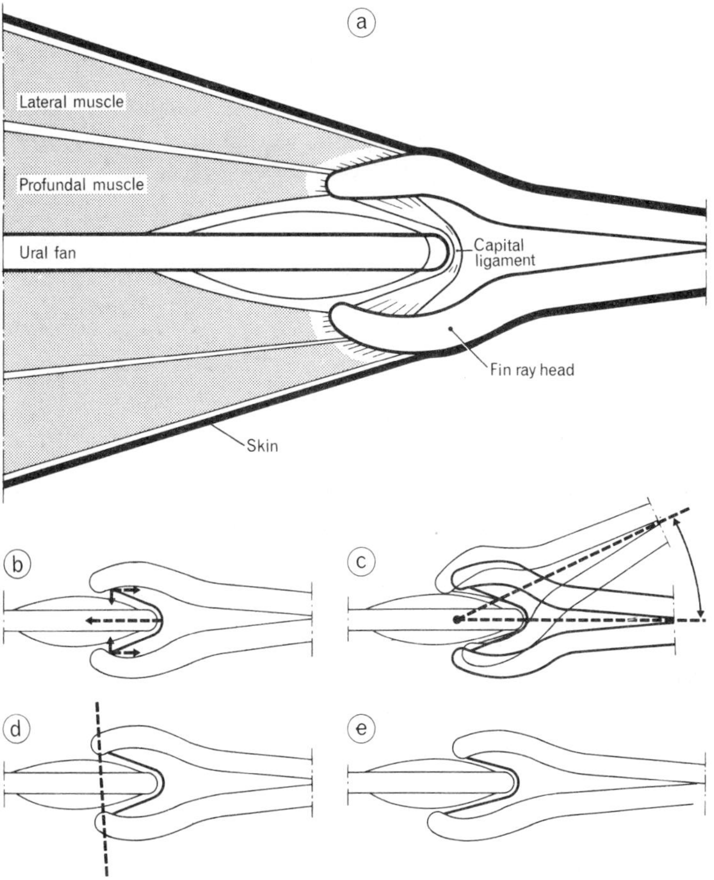

Fig. 3.12 Structure and mechanical properties of the tail fin joint of tilapia. (a) Diagram of bone, muscles and connective tissue in a horizontal (frontal) section. (b) Directions of forces caused by the pushing fin. (c) Potential for rotation in the horizontal (frontal) plane. (d) Axis of rotation for spreading and closing the fin. (e) Relative shift between hemitrichia causes the fin ray to bend or increases its stiffness against pressure from the water. Reproduced with permission from Videler (1977).

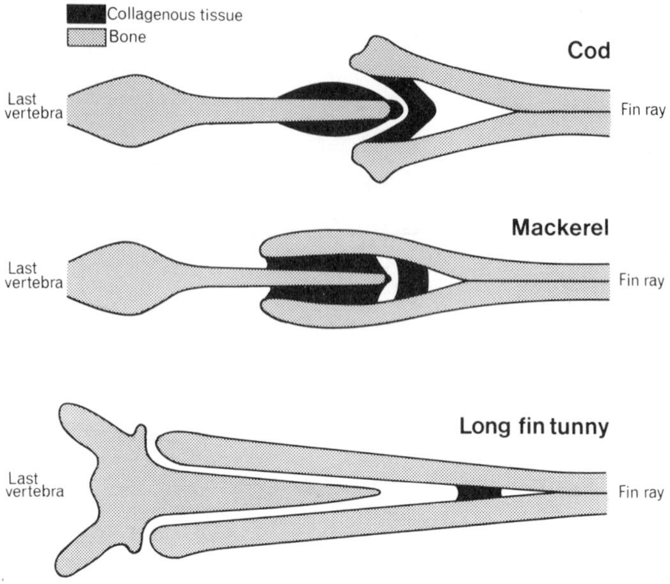

Fig. 3.13 Schematic horizontal (frontal) sections through the connection between fin and caudal peduncle of cod, mackerel and long fin tunny. Mechanically important structures are shown. Stippled areas are bone; black areas are collagenous tissue. Reproduced with permission from Videler (1981).

apart when the tail is pushing hard. The construction in the tunny is obviously far less adjustable than that of mackerel and extremely so if compared with the flexible connection of, for example, cod and tilapia.

The other unpaired fins

The dorsal and anal fins are folds of skin, usually supported by segmented skeletal structures and more or less movable owing to segmental intrinsic muscles. There is an enormous variety in forms and numbers of fins. Single elongated dorsal and anal fins are considered primitive and rows of (or single separate) fins, more advanced.

The fin ray bearers or pterygiophores consist of three pieces of bone or cartilage, the basal, middle and distal radials. The middle one tends to disappear or to coalesce with the basal radial. The basal radials are part of the median septum. Sheets of collagen connect them with the neural or haemal spines. There is in some cases considerable overlap between these spines and the pterygiophores. There is usually more than one pterygiophore per vertebral segment (see e.g. cod in Fig. 3.8).

The supporting elements of the selachian fins are cartilaginous. The pterygiophores consist of three or two elongated cartilaginous radials. The distal radials are at the basis of the fins between two sheets of densely packed ceratotrichia. The fin area of elasmobranchs cannot be altered owing to the lack of space between the ceratotrichia. The only movement that a one-sided contraction of the intrinsic muscles can cause is lateral deflection of the fin out of the median plane.

The basic design of the fins of bony fishes offers the possibility of far greater mobility. There is only one ray in each segment, and the rays are connected by a webbed flexible connection leaving a lot of space between the rays. This construction allows opening and closing like fans and lateral movements of larger amplitude and shorter wavelength. This becomes clear if we take a detailed look at the dorsal fin of tilapia (Geerlink and Videler, 1974), serving as a typical example for teleost anal and dorsal fins.

The drawing of the skeleton of this fin in Fig. 3.14 clearly shows that it consists of two parts. The first 16 or 17 fin rays are stiff unsegmented spines. The posterior part is supported by between 11 and 15 flexible segmented rays. (The numbers vary somewhat between individuals.) Spines are stiff unsegmented fin rays, with the fin ray halves united into one firm pointed structure. The flexible or soft rays strongly resemble the segmented rays described in Section 3.2. Soft rays and spines articulate in a joint at the distal end of the pterygiophores. Each pterygiophore consists of a basal and a distal radial. The joints of the spines and soft rays are completely different. In the joint of the spine, drawn in perspective in Fig. 3.15((a) and (b)), the distal radial is tightly connected to the corresponding basal radial, and the top part of the next basal radial. Both carry a hook-shaped extension, pointing forward from the basal and rearward from the distal radial. These extensions fit into two sockets in the spine base. This construc-

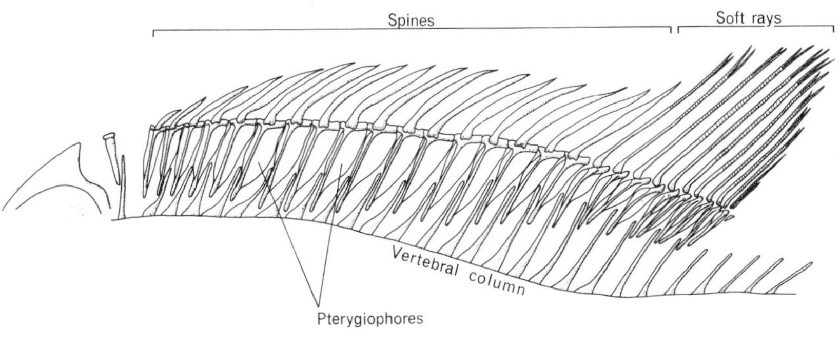

Fig. 3.14 Skeleton of the dorsal fin of the cichlid fish *Sarotherodon niloticus*. Reproduced with permission from Geerlink and Videler (1974).

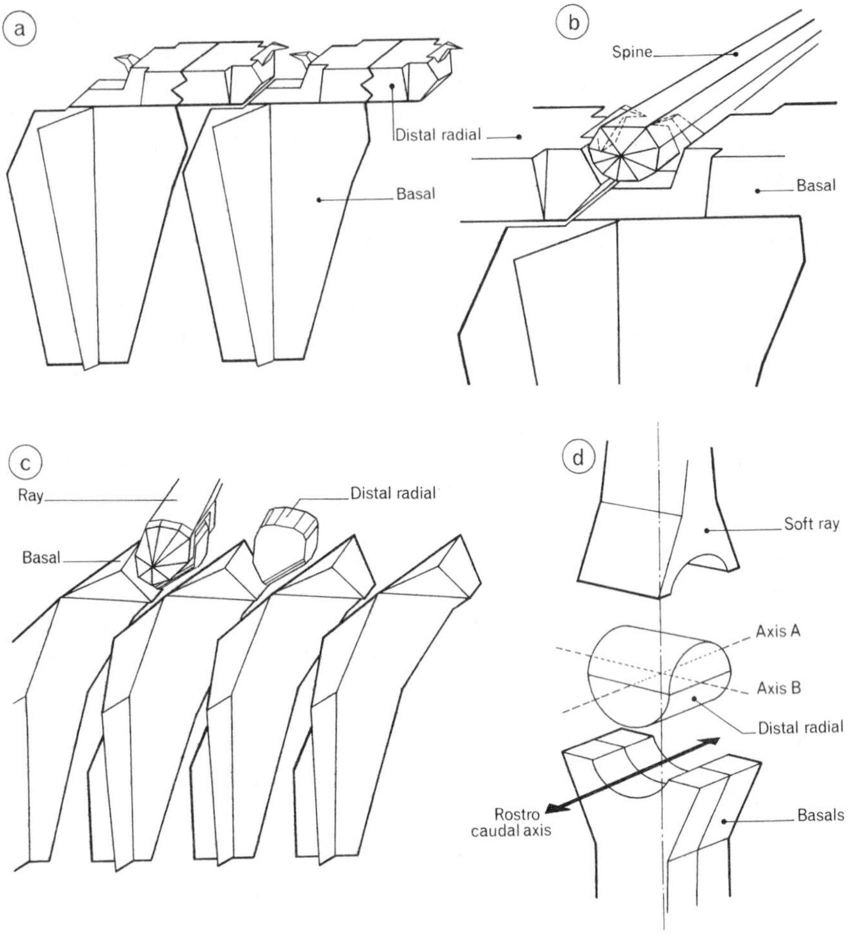

Fig. 3.15 Diagrams of the skeletal parts of a spine ((a) and (b)) and a soft ray ((c) and (d)) joint of tilapia: (a) reconstruction of the supporting elements without the spine, (b) details of form and function of the connection between spine and supporting skeleton, (c) the ball-and-socket joint of the soft rays, (d) the two perpendicular axes of movement of the soft ray joint. Reproduced with permission from Geerlink and Videler (1974).

tion prevents lateral movements of the spine, but allows rotation in the median plane.

The soft rays have a kind of ball-and-socket joint, formed by two successive basal radials, one distal radial and the base of the fin ray. Fig. 3.15(c) shows these skeletal elements in perspective and Fig. 3.15(d) provides a diagram explaining the motions allowed by this type of joint. Axis of movement A, in rostro-caudal direction, allows lateral rotation of the fin ray to both sides. The upper part of the ball-shaped distal radial makes rotations in the median plane around axis B possible. Furthermore, the stiffness or the lateral camber of the soft rays can be altered by a relative shift of the fin ray halves (page 48).

At the base of each ray we find a common pattern of muscle insertion. Two erector muscles, one at each side of the median plane, and likewise two depressor muscles, originating from the basal radials and the median septum, cause rotation in the median plane. The erector and depressor muscles insert on the rostral and caudal side of the fin ray base respectively. Inclinatory muscles, pulling the ray sideways, originate from the inner side of the skin and insert on the lateral ray base on each side. In case of the soft rays, these muscles are larger and have a more expanded region of insertion. Separate inclinatory muscles insert on the distal and basal radials between the joints of the spines. These probably provide support for the pterygiophores of the spines when strong lateral water pressures are exerted on the erected fin.

The Scombridae possess rows of finlets behind the second dorsal and anal fins. Each finlet of a mackerel is one single branched fin ray of complex shape. The frontmost branch of each of the two fin ray halves is spiny and stiff, consisting of a few longitudinal segments only. The more rearward branches of the ray are more flexible and consist of more segments. The caudalmost branch is the thinnest and the most flexible. The pterygiophores of the finlets consist of a basal, a middle and a distal radial. The basal radial is firmly connected by collagenous fibres to the neural and haemal spines. The middle radial is a separate structure and the finlets are spaced out owing to its elongated shape. The distal radial forms a saddle-shaped joint with the base of the fin ray. Movements of the finlets are under muscular control but their function is still unclear.

The same holds true for the function of the adipose dorsal fins. Despite the name, these fins are not filled with fat. Adipose fins consist of folds of skin with connective tissue in between. They are not covered by scales and there are, with some exceptions, no fin rays nor is there a supporting skeleton. This peculiar type of fin occurs in salmonids, characoids and scopelomorph fishes as in many species of catfish (Siluriformes).

Paired fin structure and function

The shape, the size and the position on the body of pectoral and pelvic fins vary considerably among fish. I will start with a brief overview of the functional morphology of the paired appendages confined to pelagic roundfish, including sharks and teleosts. Subsequently the functional morphology of the teleost pectoral fin will get special attention. This is done because teleost pectorals are abundantly used for propulsion; in many species they are the main propulsive device. The pectoral fin structure of the wrasse, *Coris formosa*, a typical pectoral fin swimmer, will be compared with that of tilapia, a cichlid which uses the pectorals for auxiliary locomotive functions.

Paired elasmobranch fins are supported by cartilaginous plates of the pectoral and pelvic girdles. The skeleton of the fins is attached to these girdles and consists of rows of basal and radial plates. The distal part of the fins is supported by densely packed fin rays, the ceratotrichia. This type of skeletal support makes rather rigid fins. Intrinsic fin muscles are presumably able to change the camber a little bit, but there is no way to alter the fin areas. The tolerances in the connections between the cartilaginous elements enable slight, muscle-induced rotations of the fins with respect to the girdles. There is no clearly defined joint. The approximate axis of rotation in shark pectoral and pelvic fins is horizontal. The fins have a fixed area and a broad base; the pectorals are larger than the pelvics.

Sharks are usually heavier than water and the swimming thrust of the tail has, as we saw, an upward component. The head-down pitching moment caused by this vertical force is counteracted by lift generated by the wing-shaped pectoral fins. These make a small (usually less than $8°$) angle of incidence with the horizontal (Harris, 1936). The pectorals are positioned in front of, and the pelvics behind, the centre of mass. Upward forces generated by the pelvic fins cause a small head-down pitching moment which enhances the effect of the epicercal tail. Swimming sharks are statically unstable which increases their controllability (Harris, 1936). The ventral position and the wing-like appearance of pectorals and pelvics help to prevent rolling.

We saw that the thrust generated by the homocercal tails of neutrally buoyant teleosts is directed horizontally. Neutral buoyancy eliminates the necessity for the generation of lift forces. Scombridae, however, are teleosts heavier than water. They normally use their pectorals to generate dynamic lift (Magnuson, 1973). At very low swimming speeds these dynamic lift forces are insufficient to avoid sinking; mackerel are observed to swim with their body tilted upwards under these circumstances (He and Wardle, 1986). The tilted body provides some extra lift, and swimming forces from the tail are no longer directed horizontally but slightly upwards along the direction of the vertebral column. Under these circumstances the tail imitates the epicercal movements of shark tail fins.

Teleost fins are highly mobile, collapsible and capable of changing their effective area. The position on the body of the pectorals and pelvics varies among teleosts. Disregarding a fair number of exceptions, one could distinguish three types of positioning. Fig. 3.16(a) shows the shark-like configuration with the pectoral fins low up front and the pelvics far rearward. It is

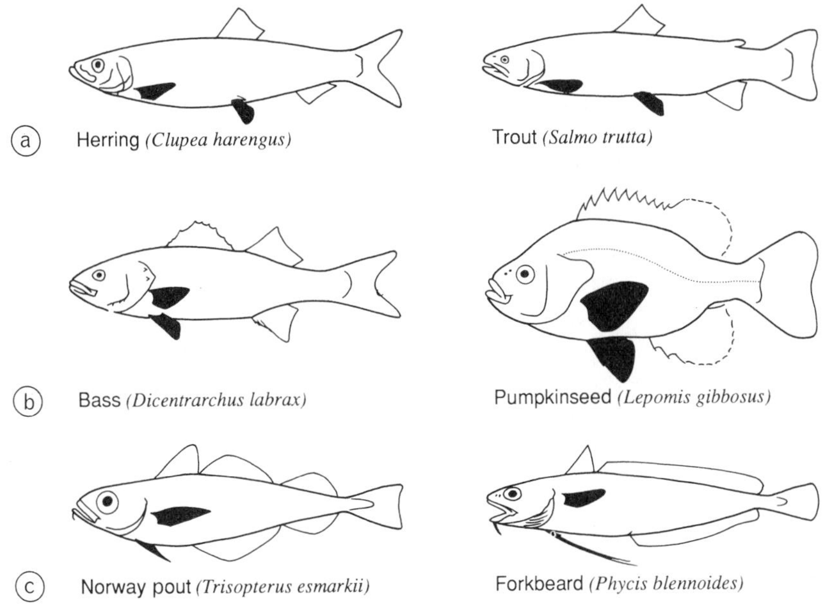

Fig. 3.16 Relative position of pectoral and pelvic fins in representatives of the three configurations that can be distinguished. (a) The shark-like positions: pectorals low up front and pelvics rearward (e.g. herring, *Clupea harengus*, and trout, *Salmo trutta*). (b) The positions found in most advanced teleosts: pectorals high on the body above the pelvics (e.g. bass, *Dicentrarchus labrax*, and pumpkinseed, *Lepomis gibbosus*). (c) The gadoid configuration: pectoral fins high on the body behind the pelvics (e.g. Norway pout, *Trisopterus esmarkii*, and forkbeard, *Phycis blennoides*). Modified after Harder (1975b).

found among sharks, sturgeons and teleost groups that are regarded as more primitive, such as salmonids, catfish, pikes, cyprinids and herrings. The most common situation for the more advanced perciforms is represented by Fig. 3.16(b). The pectoral and pelvic fins are at the same longitudinal position and the pectorals are situated higher up. The gadoids form a separate group because the pelvics are situated in front of the pectorals (Fig. 3.16(c)).

The pelvic girdle of teleosts is not connected to the axial skeleton. The skeleton that bears the fin rays consists usually of two triangular bony elements. The longest sides of the triangles are connected in the median plane along a bony symphysis. The rearward base of each triangle is shorter than its length. The forward-pointing tips of the plates are in some cases connected with the pectoral girdle. Fin rays are attached to the bases of the plates. The connection is similar to that in the caudal fin. If spiny rays are present, these are positioned on the lateral sides and soft rays are found in the middle. Various groups of intrinsic dorsal and ventral muscles originate on the triangular plates and insert on the fin ray heads. (See Baerends and Baerends-van Roon, 1950, for a detailed example of the muscular system of tilapia.) The dorsal muscles on the inside are the adductors, pulling the fins close to the body and folding the fin rays together. The abductors on the ventral outside rotate the fins forward. The axis of rotation is usually horizontal and approximately perpendicular to the median plane. Special muscles inserting on the outer fin rays spread the fins. The complex muscle system can also stiffen the rays or change the camber by differential pulling on the fin ray heads.

The pelvic fins are used for braking and will probably generate upward force as well as head-up or head-down pitching moments depending on their position with respect to the centre of gravity. Research and evidence about these functions are lacking. Pelvics are very often adapted to a variety of functions not related to swimming.

The shoulder girdle and fin on the left and right side of the body of a typical teleost are mirror images of each other. Tilapia (Geerlink, 1979) is used once again as an example of a typical teleost and its structures are drawn in Fig. 3.17. The skeleton of one side is commonly divided into four groups of elements.

1. The secondary arch consists of four pieces of bone, the posttemporal, supracleithrum, cleithrum and postcleithrum. There are two connections with the skull. The posttemporal is attached to the upper part of the neurocranium, and Baudelot's ligament connects the ventrocaudal side of the neurocranium with the inner side of the tip of the cleithrum. Ventrally the left and right cleithrum are connected, closing the girdle. The cleithrum is the largest piece and the other elements are attached to it.
2. The primary arch consists of scapula and coracoid. These are flat plate-like structures, each with a foramen and reinforced with thickened ridges.
3. There are four proximal radials and a fibrocartilage pad containing the remnants of 12 distal radials.
4. There are 15 fin rays, each consisting of two hemitrichia. The connection between the primary and secondary arch is firm, and the proximal radials

---

Fig. 3.17 Anatomy of the shoulder girdle and pectoral fin of *Sarotherodon niloticus* (Cichlidae). (a) Lateral and caudal view of the skeleton of the shoulder girdle. (b) The cleithrum and the primary shoulder girdle. (c) Schematic representation of the joints between the fin rays and the proximal radials. (d) The shapes of the heads of the hemitrichia. Abbreviations: Bl, Baudelot's ligament; cl, cleithrum; cor, coracoid; fcp, fibrocartilage pad; f.gl, fossa glenoidalis; lh, lateral hemitrichia; mh, medial hemitrichia; pcl, postcleithrum; pr.rad, proximal radials; pt, posttemporal; scap, scapula; scl, supracleithrum. Reproduced with permission from Geerlink (1979).

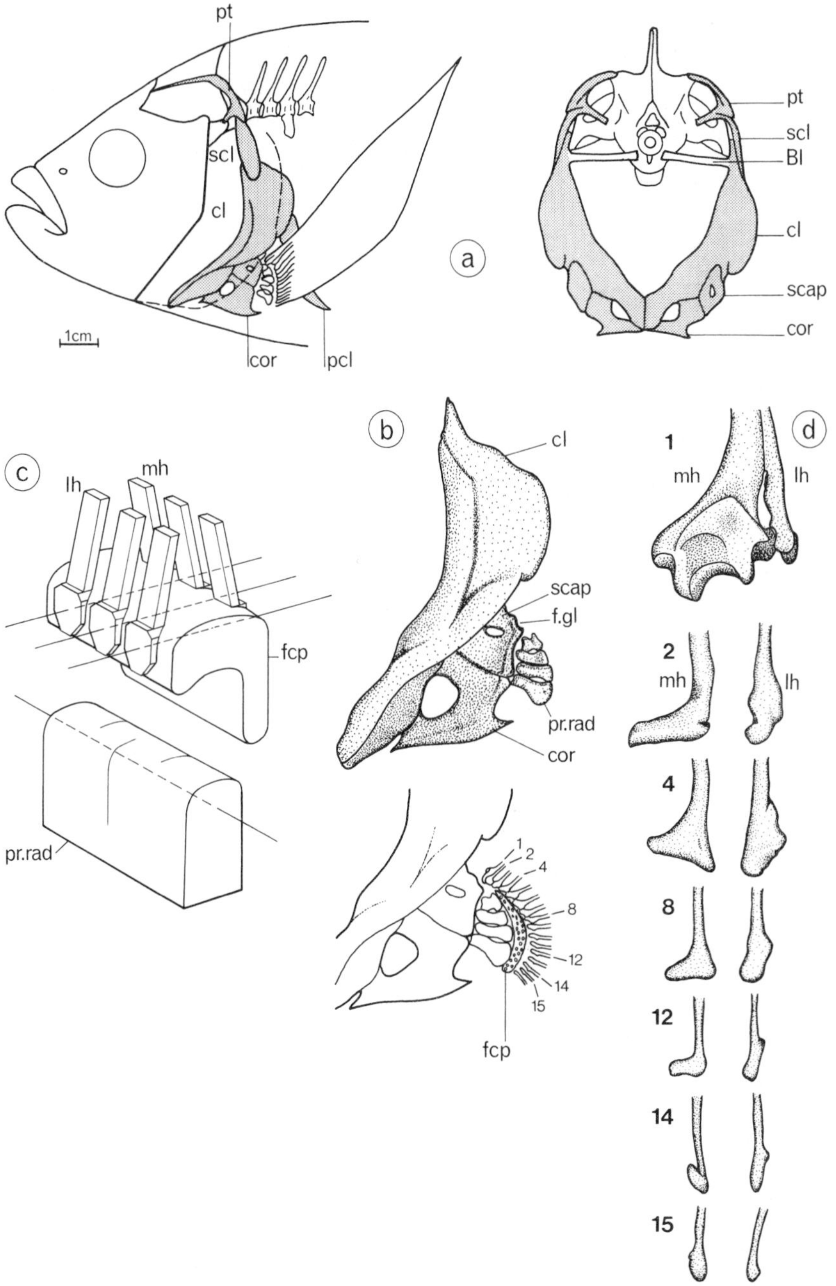
pt
scl
cl
a
1cm
cor
pcl
pt
scl
Bl
cl
scap
cor
c
lh
mh
fcp
pr.rad
b
cl
scap
f.gl
pr.rad
cor
1
2
4
8
12
14
15
fcp
d
1
mh
lh
2
mh
lh
4
8
12
14
15

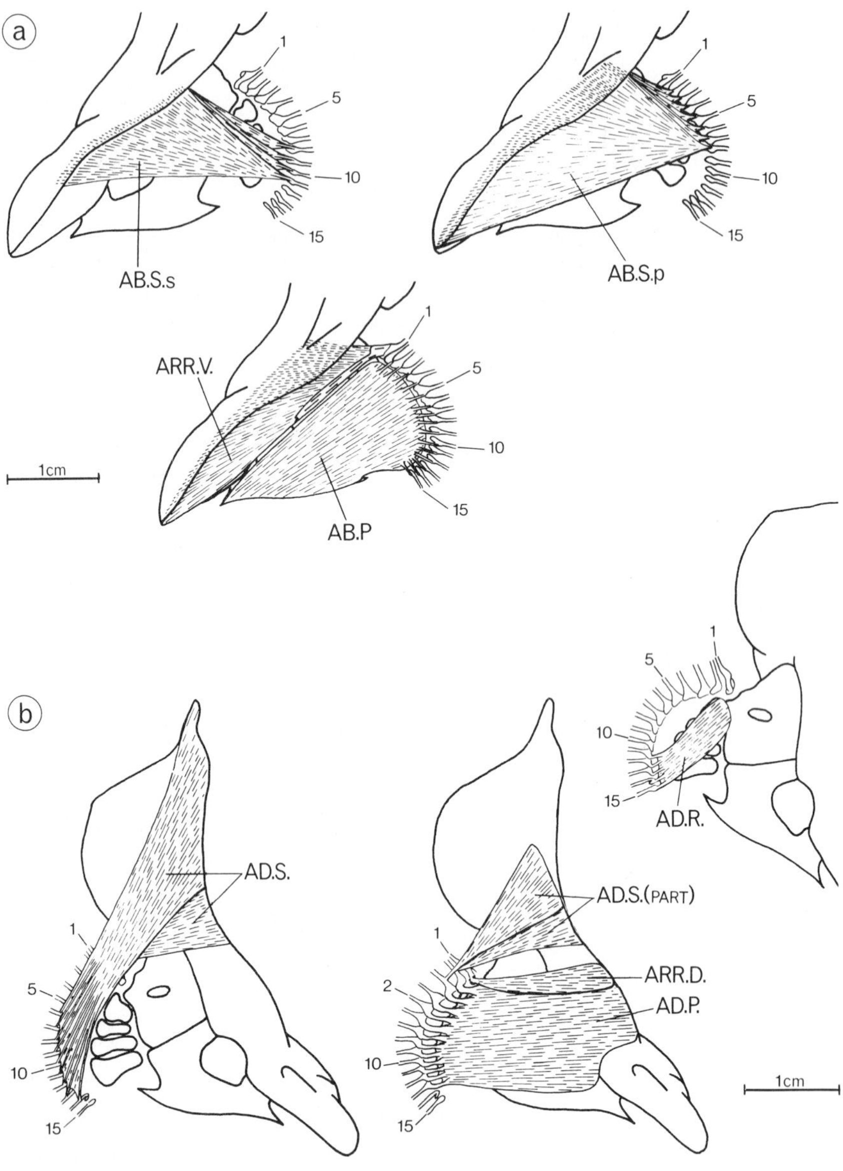

Fig. 3.18 The abductor (a) and adductor (b) muscles of the pectoral fin of tilapia. Abbreviations: AB.P, abductor profundus; AB.S.p, abductor superficialis, profundal part; AB.S.s, abductor superficialis, superficial part; AD.S., adductor superficialis; AD.P., adductor profundus; AD.R., adductor radialis; ARR.D., arrector dorsalis; ARR.V., arrector ventralis. Reproduced with permission from Geerlink (1979).

articulate with the caudal ridge of the scapula and coracoid. The distal ends of these radials are covered by the fibrocartilage pad. It covers more of the radials on the medial side than on the lateral side. Microscopic sections through this structure reveal the presence of cartilaginous spherical bodies, the distal radials. The bases of all but the first fin ray (counting downwards) are partly embedded in the pad. The first fin ray is rather smaller than the second one and is strongly connected to the second over its full length. The shapes of the fin ray heads are reminiscent of the various shapes of golf clubs. The knobby ends differ between the rays but also between the medial and lateral hemitrichia of the same ray. The fin ray length increases steeply from the combined first and second to the fifth, and declines from that one gradually towards the fifteenth.

The range of possible fin movements is virtually endless: tilapia is able to control every fin ray separately; the fish can spread and close the fin and rotate the rays forward and backward, as well as up and down. It also controls the individual camber of the rays by shifting the hemitrichia along each other. The muscular system powering these versatile movements is rather complex. There are four abductor muscles with different origins and insertions on the lateral fin ray heads (Fig. 3.18(a)), and an equal number of adductors on the median side (Fig. 3.18(b)). Small differences between insertions as well as multiple insertions on different positions on the shafts and knobby ends of the rays make very complex multidirectional movements possible. Geerlink (1989) compared pectoral fin morphology and motion patterns of the cichlid tilapia with those of two species of wrasse which use the pectorals predominantly for unidirectional swimming movements. Surprisingly, detailed morphological comparisons showed only minute differences that could be related to the different movement patterns between the cichlid and the wrasses. Geerlink's conclusion was that the different movement patterns are mainly based upon differences on the cybernetic level. Studies of motor patterns are required to test this.

3.5 SUMMARY AND CONCLUSIONS

The median septum dividing the body of fish into two lateral halves is supported by the vertebral column between head and tail. The number of vertebrae differs considerably, not only between species but also intraspecifically. Vertebrae are fitted with neural and haemal arches and spines. Collagenous intervertebral connections, elastic dorsal and ventral ligaments, threaded through the neural and haemal arches, and the backwardly inclined spines, stayed by the collagen of the septum, potentially act as springs to restore the straight attitude of the body axis.

The building elements of fish fins are the fin rays. These structures make the fins of fish extremely well adapted to the propulsive interaction with water. The basic building principle and the bending properties of a single ray are described in detail.

Fins are the limbs of fish; structure and function of caudal fins, dorsal and ventral unpaired fins and pectoral fins are described. In each case a detailed example is given, showing the basic design and details of the skeleton and the intrinsic musculature.

Chapter four

The structure of the swimming apparatus: shape, skin and special adaptations

4.1 INTRODUCTION

The shape of the swimming body should ideally be streamlined, offering the lowest drag for the largest volume. Nevertheless, the shapes of fish bodies vary considerably, and we want to find out why.

Fish skin usually consists of a thick layer of collagenous fibres with scales and mucus on the outside. The skin covers the body and pliantly conforms to its undulations, but that is probably not its only function in locomotion. We need to know more structural details before speculation about other functions turns into a useful exercise.

Electron-microscopic pictures are used to discover the layered structure of the mechanically important crisscross collagenous fibres of the stratum compactum. Existing knowledge about the mechanical properties of this part of the skin will be reviewed.

Scales are the fish's armour, yet some of the variation in the shape and structure of scales can be related to locomotion. There are many different sorts of scales. This variation may be based on fundamentally different embryological origins between distant groups of fishes or may reflect different functions of scales on various parts of the body of one fish. Hypotheses about form and function of fish scales will be discussed.

The role of mucus as an agent to reduce drag is still disputed. Does mucus lubricate fish or is it only a protective shield against the infective outside world? We shall concentrate on drag reduction, and a few published measurements may help to elucidate the processes involved.

Every species is adapted to be the fittest in its niche. Some adaptations are strongly determined by special locomotory skills, whereas other fish are jacks of all trades, performing most swimming functions reasonably well. The penultimate section of this chapter categorizes some of the specialists and tries to define the structural adaptations needed to qualify for each of these special categories.

4.2 STREAMLINED BODIES

Various attempts have been made to classify the variety of different body shapes in fish (e.g. Webb, 1984b). Some shapes are clearly related to swimming capacity. Spindle-shaped, so called streamlined bodies usually belong to the fast swimmers, and the lateral compression of many diamond-shaped coral reef fishes betrays extreme manoeuvrability. Many other forms, however, may have nothing to do with swimming but are related to other traits.

Pelagic fish allocate a considerable amount of their lifetime to swimming, and are therefore expected to do that efficiently. Natural selection will have favoured shapes that help to decrease the energy needed for acceleration and sustained swimming. The savings can be used for growth and reproduction (Chapter 10). A reduction of the amount of energy needed for swimming can be achieved by a drag-reducing body shape. There are usually two important sources of drag in fish: friction drag and pressure drag. Friction drag is proportional to the surface area of an object in a flow of water. A spherical body has the smallest surface area for a given content, and offers the smallest drag attributable to friction. Pressure drag is minimal for a very thin, needle-shaped body with its longitudinal axis parallel to the flow. We are obviously looking for a hybrid between a sphere and a needle. Hertel (1966) tells a nice story about Sir George Caylay who suggested in 1800 that the body of trout should be the model for aerodynamic design in aircraft. In 1956, von Karman found that Caylay's sketch of the outline of a horizontal section through the thickest part of a trout corresponded quite well with a modern laminar flow profile. The gliding trout pushes the water gradually aside up to the thickest part (D) at about one third of its length (L) and uses two-thirds of the body length for the thickness to decrease. Rotation of a laminar profile around its longitudinal axis describes a streamlined body of rotation. These bodies with different D/L ratios have been extensively tested in wind tunnels to find the shape that gives the lowest drag for the largest volume, a configuration desired by aircraft builders as well as by fish. The lowest surface drag coefficient is found for shapes with $D/L = 0.22$, and the lowest pressure drag occurs if $D/L = 0.24$ (Hertel, 1966). These figures

Table 4.1 Approximate thickness-over-length ratios (*D/L*) of the bodies of fish and other nektonic animals

Species	D/L
Bluefin tuna (*Thunnus thynnus*)	0.28
Swordfish (*Xiphias gladius*)	0.24
White shark (*Carcharodon carcharias*)	0.26
Cod (*Gadus morhua*)	0.16
Mackerel (*Scomber scombrus*)	0.14
Eel (*Anguilla anguilla*)	0.05
Blue whale (*Balaenopterus musculus*)	0.21
Bottle-nosed dolphin (*Tursiops truncatus*)	0.25
Emperor penguin (*Aptenodytes forsteri*)	0.26
Harp seal (*Phoca groenlandica*)	0.24

are quite similar. Streamlined bodies with D/L values between 0.18 and 0.28 are within 10% of the minimum drag values possible. Table 4.1 lists values found for swimming animals. It shows that fast swimmers are properly streamlined, but also that an eel lacks the right body shape to be an efficient low-drag swimmer.

Cross sections at different positions along the body of fast-swimming fish show that the outlines are hardly ever circular. Fish deviate from bodies of rotation usually because they are laterally compressed in the mid-region and dorsoventrally flattened near the head and caudal peduncle. Houssay (1912) related this change in shape along the fish from head to tail to the shape of a jet of water issued from an elliptical orifice. Weihs (1981) shows that this similarity is a coincidence and offers an explanation relating the different shapes of the cross-sectional areas to efficient swimming. The amplitudes of the lateral movements are usually minimal near the position of the centre of gravity of a fish. From that point along the body, the amplitude increases slightly towards the head and substantially in the direction of the tail. The effort required for oscillating lateral movements of any section of the body is a function of the hydrodynamic resistance or drag and of the inertia of the mass of water affected by the movement. We saw in Chapter 1 that the virtual or added mass is proportional to the square of the largest dimension perpendicular to the direction of motion. This implies that sections A and C, near head and peduncle respectively, of the shark in Fig. 4.1 require less effort to oscillate laterally than section B for two reasons: the masses of water that need to be accelerated at the start of each lateral movement and the drag while moving are smaller for sections A and C than for section B. The flattened head facilitates turning manoeuvres. The higher depth of the central part of the body makes lateral movements more difficult. There-

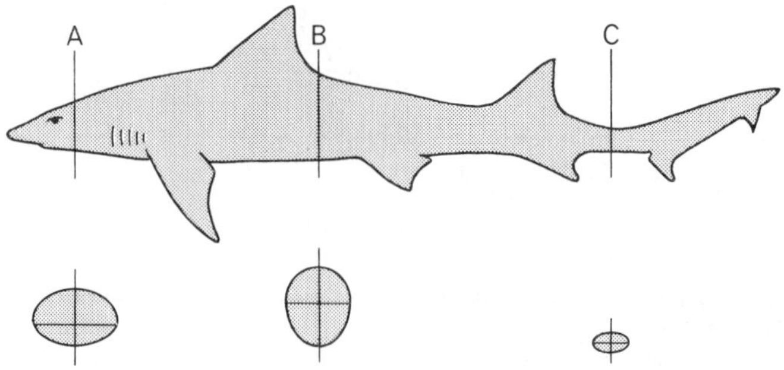

Fig. 4.1 The distribution of the shapes of oval cross sections along the body of a pelagic shark, a smooth dogfish, *Triakis acutipinnae*. Reproduced with permission from Weihs (1981).

fore, this part becomes the centre of the turn. In straight-line swimming, lateral compression of the mid-part of the body offers stability by minimizing excursions sideways. The mid sections of mackerel and tuna are only very slightly laterally compressed. These species are not specialized in sharp turns, and recoil motions are minimized in this part of the body because lateral motions are restricted to the rear one-third of the body (Weihs, 1989). The caudal peduncle with its extreme dorsoventral compression can sweep through the large amplitudes needed to drive the caudal fin. Drag arising from the lateral movements and the actual and added mass are strongly reduced to minimize the forces in the interaction between peduncle and water. The caudal peduncle of the fastest swimmers is usually provided with sharp lateral keels to reduce drag during motions from side to side. The horizontal flattening also may provide some lift in negatively buoyant species (Magnuson, 1970). The forces between fish and water should be maximal and in the right direction at the large laterally flattened tail blade. Fast swimmers among teleosts and elasmobranchs, e.g. members of the Scombridae, Xiphiidae, Istiophoridae and Lamnidae, have double keels on each side of the caudal fin. These usually converge towards the rear, probably serving to accelerate and direct water over the middle part of the fin (Aleyev, 1977).

Fast swimmers are usually equipped with a variety of structures improving the streamlined shape. Fins that are not in use during fast forward swimming fit neatly into special slots and grooves, and eyes are covered with transparent adipose tissue to keep the surface of the head flat and streamlined. Fins of the fastest swimmers are typically positioned behind the deepest part or shoulder of the body.

4.3 THE STRUCTURE OF FISH SKIN

The morphology of the integument, including skin, scales and mucus, receives special attention in a book on fish swimming because we must discuss its swimming-related functions. However, some of these are either unclear or at least sources of controversy. Bullock and Roberts (1974) treated fish dermatology thoroughly, so here only locomotion-related structures are selected. I want to separate scales and mucus from the rest of the integument because these are, as far as swimming is concerned, mainly involved in drag reduction. In many species, the leathery part of the integument is strong enough to act as a tendon and help with the transfer of forces from the lateral muscles to the tail blade, but we don't know how. This section is restricted to a formal description of swimming-related anatomical details of typical fish (e.g. tilapia) skin. Speculations about the functions are in Chapter 8.

From outside inwards are an epidermis, a dermis and a hypodermis. The epidermis contains mucus but no blood vessels. The dermis has a well-vascularized outer layer, the stratum spongiosum, where the scales have their bases, and an inner stratum compactum built out of strong fibrous connective tissue. The hypodermis is a thin layer of connective tissue, with chromatophores and lipid cells. Fig. 4.2 provides an overview on transection. The mechanically important stratum compactum consists mainly of layers of parallel collagenous fibres (Fig. 4.3). The fibres in adjacent layers are orientated in different directions but the fibre direction in every second layer is the same, comparable with the directions of the grain in different layers of plywood. The number of layers varies between 10 and 50, and is smaller in skin covering head and fins. Bundles of radial collagenous fibres run perpendicular to the surface of the skin, connecting the layers of collagen (Fig. 4.4). The angle between the fibre directions in the alternating layers can be detected from the outside of the fish after removal of mucus, scales and pigmentation, using oblique light (Fig. 4.5). The fibres follow left- and right-hand helices over the body surface. This pattern of organization of the dermal collagen fibres seems to be the general pattern because it has been described for sharks (Motta, 1977), eel (Leonard and Summers, 1976; Hebrank, 1980), scombrids (Fauré-Fremiet, 1938), a goby, *Chasmichtys gulosus* (Fuji, 1968), the knife fish, *Notopterus notopterus* (Mittal and Banerjee, 1974) and a cichlid (Videler, 1975).

The angle between the crossing fibre layers and the longitudinal body axis varies in sharks from 90° on the head to almost 0° in the caudal fin (Motta, 1977). These extreme values do not provide a useful indication of the general pattern. Along most of the animal, the angle is between about 50° and 70°. The angles between the longitudinal axis and the collagenous fibres of tilapia decrease from about 65° near the head to on average 45° in the

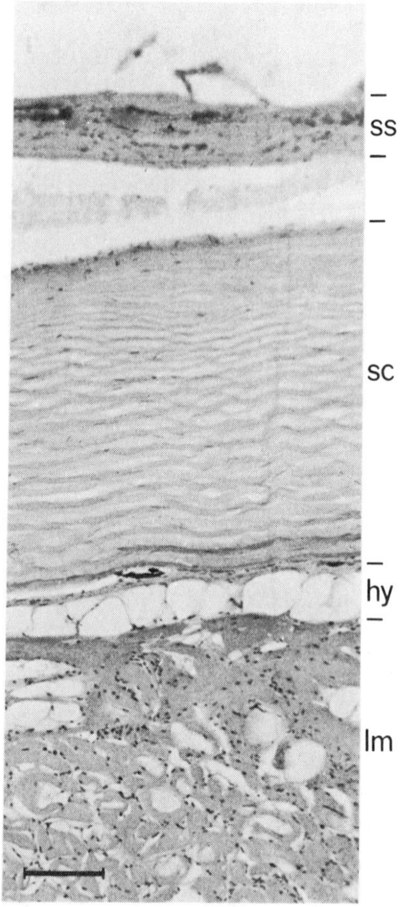

Fig. 4.2 Light-microscopic picture of a transection through the dermis (top) and hypodermis of the skin of tilapia. The stratum spongiosum (ss) and stratum compactum (sc) can easily be recognized. The hypodermis (hy) is a thin layer between the stratum compactum and the lateral muscles (lm). The scale bar is 25 μm.

caudal peduncle. The sharpest angle between the fibres is about 50° over the first one-third of the distance between head and tail and increases over the remaining part of the body up to 90° in the caudal peduncle.

The stratum compactum of eel, shark, cod, tilapia and trout is firmly attached to the myosepts of the lateral muscles but only in a narrow zone above and below the horizontal septum (Willemse, 1972; Videler, 1975; Wardle and Videler, 1980a; Ulrike Müller, personal communication,

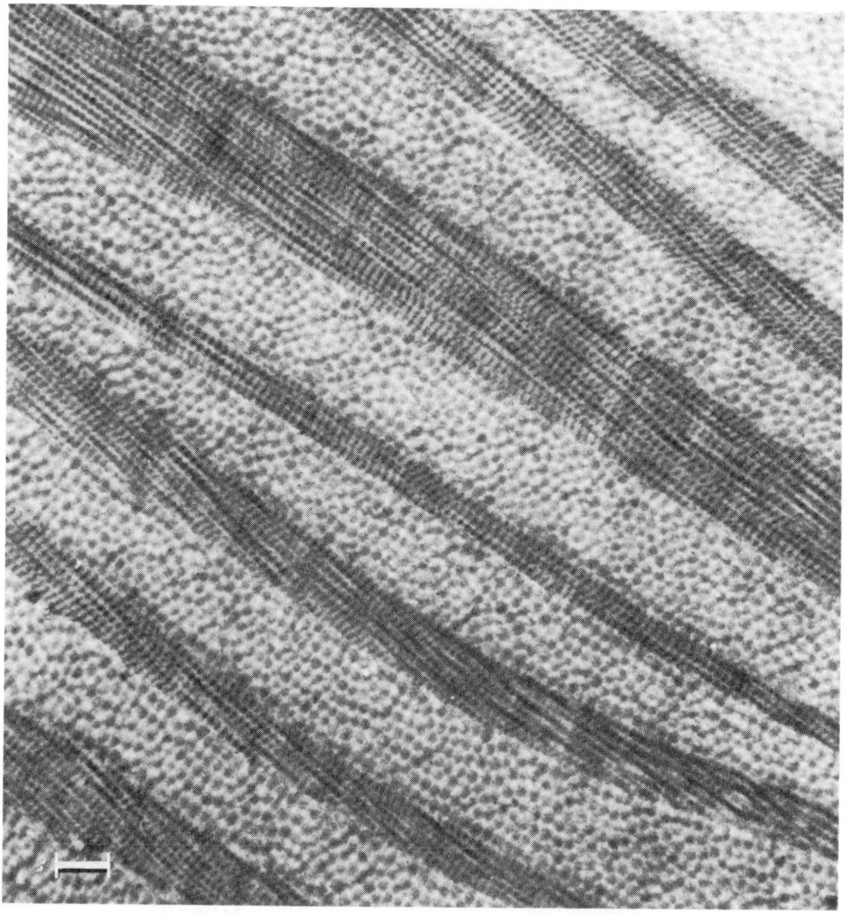

Fig. 4.3 Electron-microscopic (EM) picture of the plywood structure of the stratum compactum of the skin of tilapia. The transection shows layers of collagenous fibres orientated at approximately 90° in subsequent layers. The layers where the collagenous fibres were sectioned longitudinally show a striped pattern of alternate light and dark bands with a periodicity of 720 Å on average, typical for collagenous fibres. The layers in which the fibres are in cross section reveal that the diameter of the fibres is about 400 Å on average. The scale bar is 2700 Å.

1992). This zone coincides with the area occupied by the red muscle fibres (Fig. 2.3). In contrast, Hebrank and Hebrank (1986) did not find such firm connections in Norfolk spot, *Leiostomus xanthurus*, and the skipjack tuna, *Katsuwonus pelamis*. These contradicting observations point in the direction of real differences between species. This opinion is confirmed by fishmongers

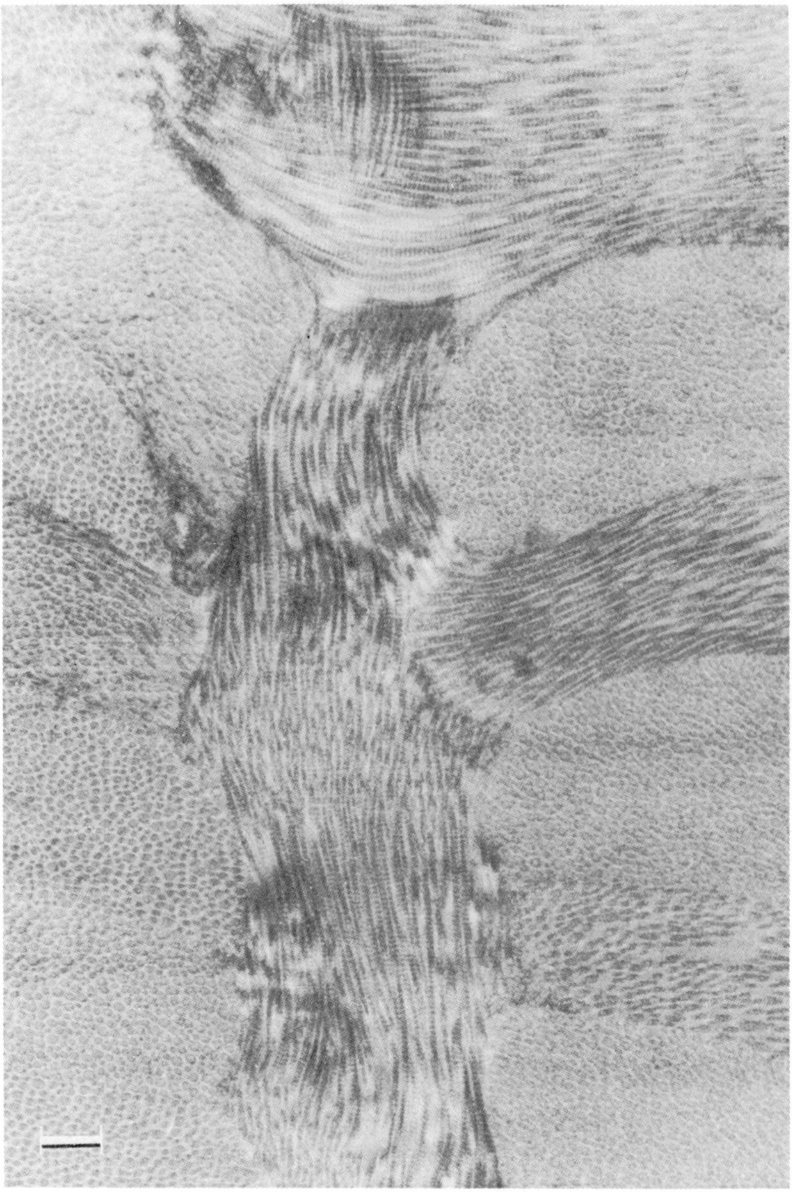

Fig. 4.4 EM picture of a radial bundle of collagenous fibres, running across the layers of crossing fibres of the stratum compactum, perpendicular to the surface of the fish. The scale bar is 1 μm.

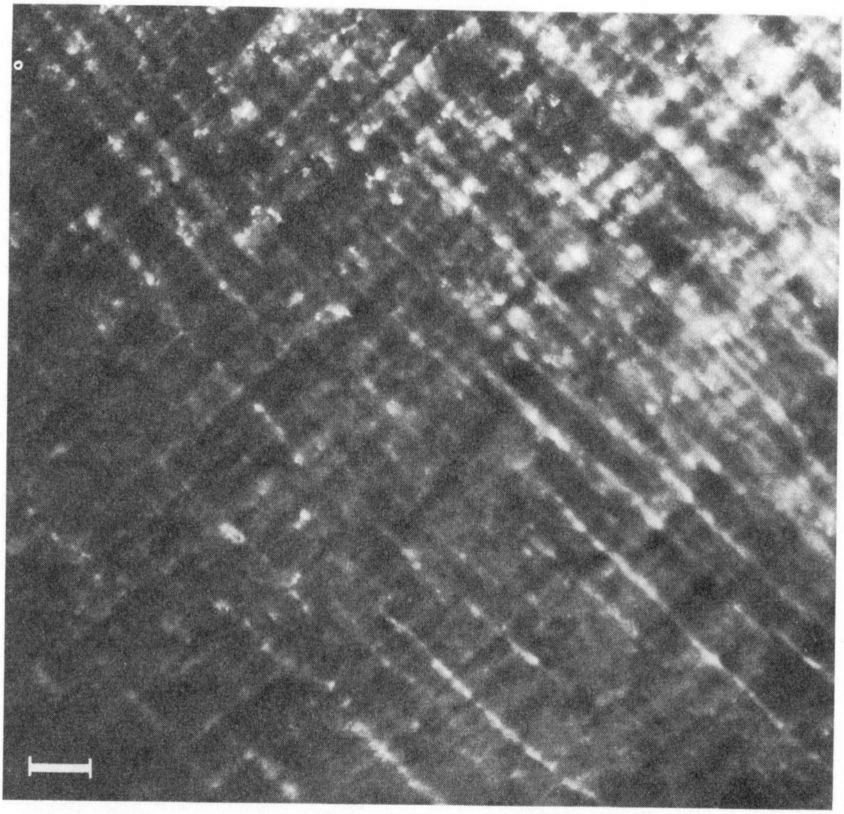

Fig. 4.5 The fibre directions in the layers of crossing fibres in the skin of tilapia can be detected on pictures of the surface of the skin taken in oblique light . The scales and mucus have been removed. The angle between the fibre directions is approximately 90°. The scale bar is 0.01 mm. Modified from Videler (1975).

who told me that some species are more difficult to skin than others. A more systematic study is obviously required.

The skin of some fish species is extremely strong. Dutch farmers used eel skin to connect the heavy swingle of a flail to its wooden handle, and eel skin has been in use as door hinges in Scandinavia. Measurements show, however, that there is a large variation in the strength of fish skin between species. Strength measurements are usually carried out by pulling at a piece of material, fixed between two clamps, while the force applied and the change in (relative) length are measured. The result of this experiment is a stress/strain curve. The stress is expressed in $N\,m^{-2}$ and strain is the percentage length change. Tendon (as it is found in the ligamentum nuchae of large herbivorous mammals) initially shows a straight relationship when stress increases linearly with strain, until it breaks at about 8% strain (Alexander, 1988). The slope of the stress/strain curve just before breaking occurs is used as a measure of strength and is known as Young's modulus (E). E is in fact the stress in Pa (1 Pa = 1 $\mathrm{N\,m^{-2}}$) needed theoretically to double the length of the object tested. The maximum value for E of vertebrate tendon is in the order of 1.5 G Pa (= 1.5×10^9 Pa). Hebrank (1980) and Hebrank and Hebrank (1986) tested the skin of eel, shark, Norfolk spot and skipjack tuna. The highest E values were found when the skin was tested in one of the two fibre directions. The values for eel and shark were 0.16 and 0.43 GPa respectively. This is almost one order of magnitude weaker than proper tendon, but one order of magnitude stronger than the skin of Norfolk spot and skipjack tuna. The E value of Norfolk spot was 0.075 GPa and that of the skipjack 0.036 GPa. So once again these data point in the direction of a dichotomy among fish species regarding skin properties. The Hebranks did not manage to measure the actual breaking strain, because the pieces of skin failed at the clips attaching the skin to the testing machine. Müller *et al.* (1991) measured the strain along the fibres in a swimming trout with strain transducers attached to the skin. The strain oscillated in harmony with the undulations of the body and increased up to 3% for swimming speeds up to 5 body lengths $\mathrm{s^{-1}}$. Assuming a Young's modulus of 0.04 GPa, the stresses occurring in the fibre direction of a trout swimming at 5 $L\,\mathrm{s^{-1}}$ would be maximally in the order of 0.001 GPa.

The conclusion of this brief survey is that the stratum compactum in fish skin varies between species in thickness, strength and with regard to the degree of connection with the lateral musculature.

Form and function of scales and mucus

Fish scales

Fossil records reveal that ancient fish, e.g. the placoderms, were armoured with bony plates of a dermal skeleton. This probably restricted their mobility. It is tempting to imagine that scales are an adaptation offering greater agility with higher swimming speeds, without losing the protection of armour altogether. Lampreys and hagfishes, the recent representatives of the ancient group of jawless fishes, have no scales or remains of a dermal skeleton whatsoever. There are also teleost families with very small scales or none at all. Scales are reduced or absent on the rear part of the fastest teleosts. However, these are the exceptions, and in general fish are covered with scales arranged in regular patterns. There are many different types, and their dissimilar histology and embryonic development provided arguments for the reconstruction of evolutionary pathways (Harder, 1975a, provides an extensive and Alexander, 1981, a practical overview).

Elasmobranchs possess placoid scales, constructed in essence as vertebrate teeth. Placoid scales have a basis of bone with a central pulp cavity, are covered with dentine and are equipped with a crown of hard enamel-like material. Once formed, these scales do not grow. They are repeatedly shed and their number increases when the animal grows.

Teleost scales are, in contrast, simple plates of bone, growing continuously at the outside. Daily growth increments can usually be detected and seasonal differences in growth rates also show up in every scale. This is the elasmoid type of scale. It is called cycloid when round and ctenoid when the edge bears comb-like projections.

Teleost scales usually strongly overlap, but the ganoid scales of bichirs (Polypteridae) and gars (Lepisosteidae) do not. These are very thick, made out of bone, dentine and thick layers of enamel-like material. Cosmoid scales of lungfishes (Dipneusti) and coelacanths (Crossopterygii) have four layers: two layers of bone (one deeper and a surface layer), one dentine and an enamel-like layer. Sturgeons have a rudimentary scale covering, consisting of bony plates.

Aleyev (1977) reviews a large body of data on the hydrodynamic functions of fish scales published in Russian. He provides substantial circumstantial evidence for three main functions, but without offering real proof for any of these. The first hypothetical function is related to streamlining and the second and the third are connected with control of the flow in the boundary layer.

1. Scales serve to prevent transverse folds on the side of the fish. Support for this function comes from the fact that during ontogeny, scales initially appear on the sides of the body and that scales usually disappear in fishes with insignificant body undulations.
2. Ctenoid structures, e.g. spines, dents and tubercles, are arranged to form longitudinally orientated crescents or run-off grooves serving to conduct the waterflow in a laminar fashion and prevent large-scale turbulence.
3. Microrelief on scales generates microturbulence which supposedly reduces the drag over a critical *Re* range by delaying the transition from laminar to turbulent flow.

There is ample evidence for a direct relation between the division of cycloid and ctenoid scales and the *Re* number (calculated for the typical speed of a fish of a certain length). For example, the development of scales during ontogeny of the mullet, *Mugil saliens*, shows that at *Re* below 10^3, the juvenile fish has no scales at all. Between 1 cm and 3 cm body length cycloid scales develop all over the body as the *Re* increases up to 10^4. Growth up to 10 cm and *Re* increase to 3×10^5 coincides with a gradual replacement of cycloid by ctenoid scales on the rear part of the body behind the head. At *Re* above 10^6 the ctenoid apparatus of the mullets begins to regress and at *Re* above 5×10^6 all the scales are cycloid again.

The explanation given for this phenomenon is that small fry with cycloid scales maintain laminar flow over the entire body at low *Re* numbers. At higher *Re* the flow is in danger of suddenly becoming turbulent with large drag costs. The ctenoid scales are supposed to prevent this by creating microturbulence which keeps the flow over the body in effect laminar. At *Re* numbers beyond the transition zone this trick is no longer needed and the scales can be cycloid again.

Measurements of the drag coefficient C_{dw} (Equation 1.12) of streamlined bodies, collected by Hoerner (1965), are related to *Re* in Fig. 4.6. With *Re* increasing, C_{dw} decreases gradually under laminar flow conditions and increases rather suddenly during the transition from laminar to turbulent flow. At even higher *Re* the drag coefficient decreases again, but at a lower rate than under laminar flow. It would therefore first of all pay to avoid having such a sudden transition and secondly it would be advantageous if the laminar rate of decrease could be maintained at high *Re* numbers. The creation of a microturbulent boundary layer, well before transition takes place, could possibly generate these effects. Unfortunately there is no proof yet that this happens. Detailed flow visualization techniques are needed to provide evidence.

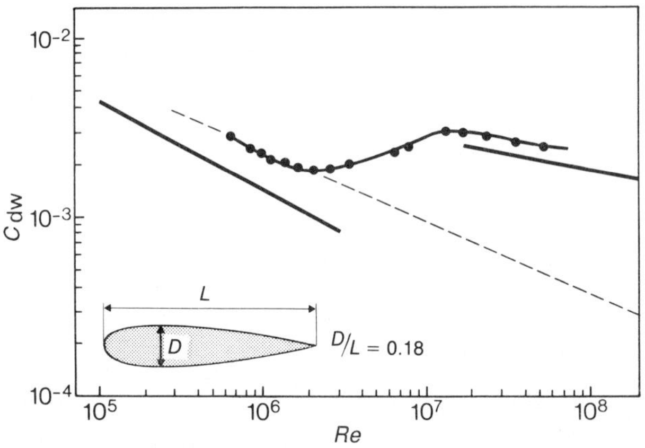

Fig. 4.6 Flow-channel measurements of the drag coefficient C_{dw} (Equation 1.12) of a streamlined, rotationally symmetric body as a function of *Re*. *D*/*L* of the body is 0.18. The lines are indications of the drag over a flat plate with a surface area equivalent to that of the streamlined body, under laminar (solid curve at left) and turbulent (solid curve at right) conditions. The broken curve suggests extrapolation of the laminar values beyond the laminar-to-turbulent transition point. Based on Hoerner (1965), fig. 22, pp. 6–16.

There are many other remarkable tiny details of fish scales that have not been functionally explained. Bone (1975) shows how the extension from the body surface of the denticles on the scales of the dogfish varies between 0.05 and 0.12 mm depending on the position on the body. The castor oil fish, *Ruvettus* sp., has ctenoid scales fitted with spines projecting up to 1 mm above the body surface. There are pores and canals between these scales allowing ejection of sea water on the convex side of the body during the swimming cycle. It is tempting to interpret these structures as drag-reducing devices, but proving how they function is a different matter. Mucus is similarly regarded as a drag-reducing agent, we turn now to an account of the existing knowledge.

Mucus

Goblet or mucoid cells are single-celled glands in the epidermis, producing mucopolysaccharides. The secretion is fibrous and swells up in water to form a thick layer of viscous mucus. Not all fish species have these cells, and the amount of mucus produced varies from enormous amounts (e.g. a bucketful by the hagfish, *Myxine glutinosa*) to none. Mucus has a protective function against infections and parasites, reduces the danger of damage to the skin through collisions and helps the fish to escape the grasp of predators. Mucus seals the body and prevents the exchange of ions and water (Harder, 1975a). There are many more special functions described for a variety of species, varying from the protective nightgown of parrotfishes to a foodstuff for young cichlids.

Mucus is also assumed to reduce frictional drag during locomotion. Toms observed in 1948 that small amounts of polymers added to turbulent high *Re* flow in a pipe reduced the pressure drop substantially below that of the fluid alone at the same flow rate. Physicists have called this the 'Toms effect' ever since (Lumley, 1969). Rosen and Cornford (1971) measured the pressure decrease of water in turbulent flow through a pipe, with and without various concentrations of fish mucus. The slime of the California bonito, *Sarda chiliensis*, had almost no effect, whereas at the other end of the scale, a solution with a 5% concentration of the mucus of the Pacific barracuda, *Sphyraena argentea*, gave a maximum reduction of almost 66% compared with pure ocean water. Similar reductions in pressure differences were reached with a 5% mucus solution of the halibut, *Paralichthys californicus*, and with a 10% slime solution of the nut brown cowry, *Cyprea spadicea*, a mollusc. The Toms effect was very poor for Pacific mackerel, *Scomber japonicus*, (50% reduction in a 50% solution and almost no reduction for a lower than 10% solution). Intermediate results were obtained for a number of freshwater species.

However, a demonstration of the Toms effect for fish mucus does not prove that mucus helps to decrease the frictional drag experienced by a swimming fish. To establish the effect of fish mucus on the *in situ* flow in the boundary layer, Daniel (1981) compared velocity profiles over the surface of freshly killed fish with those over wax models. He measured flow at *Re* between 2×10^3 and 2×10^4 and showed that rainbow trout, *Oncorhynchus mykiss*, mucus makes the velocity profile less steep (the gradient was 40% lower). In other words, the thickness of the boundary layer (defined in Chapter 1) of the fish was almost twice as thick as that of the smooth wax model under the same flow regime. In narrow boundary layers with steep velocity profiles momentum is transferred at a higher rate. The viscous shear stress acting on the surface of the fish is proportional to the steepness of the velocity gradient (Prandtl and Tietjens, 1934). This implies that mucus drastically decreases the frictional drag experienced by the fish. Drag owing to viscous friction is the main component of drag on a streamlined body. This leads to the conclusion that mucus plays an important role in overall drag reduction during swimming at low *Re* of at least some species of fish. It would be interesting to know if the deviating behaviour of the slime of the California bonito is related to its fast steady swimming habits. The steepness of the velocity gradient is probably not so much affected by mucus at high *Re* numbers. If that is so, the barracuda would only benefit from the mucus-induced drag reduction while cruising slowly and not during a fast dash for prey.

4.4 SPECIAL ADAPTATIONS

A variety of fish species is adapted to perform some aspect of locomotion extremely well, whereas others have a more general ability to move about and are specialized for different traits. Webb (1984a and b) divided fishes into three specialist groups and a majority of generalists. Among the generalists were salmonids, surfperches, sculpins and the bluegill, *Lepomis macrochirus*. I called tilapia a 'typical fish', which is probably synonymous for generalist. Generalists have bodies that give them moderately good performance in all the special functions. Webb recognized specialists for accelerating (pikes, barracudas, pipefishes), cruising (mackerel and tuna, pelagic sharks) and manoeuvring (butterflyfishes, coral reef and weedy-river fishes). There are, however, groups of fishes that do not fit in any of these four categories, or in other words the generalist category is too diverse and contains other specialist groups. At least seven other distinct swimming guilds can easily be recognized. I will briefly discuss characteristic aspects of the functional morphology and Fig. 4.7 shows typical representatives of each of these 11 expert groups.

Specialists for accelerating. The pike and the barracuda are ambush predators. They remain stationary or swim very slowly until a potential prey occurs within striking distance. Colour patterns adapted to each specific environment provide the essential camouflage to make them hardly visible against the background. Both species have a reasonably streamlined body and large dorsal and anal fins positioned extremely rearward, close to the caudal fin. Acceleration during the strike is caused by the first two beats of the tail, which is in effect enlarged by the rearward position of the dorsal and anal fins. The relative skin mass of the pike is reduced, compared with other fish, increasing the relative amount of muscles and decreasing the dead mass that has to be accelerated with the fish at each strike (Webb and Skadsen, 1979). Detailed kinematic descriptions of this behaviour are given in Chapter 6.

Cruising specialists. Scombridae and pelagic sharks with extremely streamlined bodies, narrow caudal peduncles with keels and high AR tails migrate over long distances, swimming continuously at a fair speed. The bluefin tuna, *Thunnus thynnus*, for example, crosses the Atlantic from the Caribbean to reach Gibraltar in spring. These fish enter the Mediterranean and spawn near Sicily. During the summer the spent blue-fins move out of the nutrient-poor Mediterranean and migrate north and west to the rich areas near Nova Scotia. Later in the year they move back to the Caribbean before crossing the Atlantic again.

Manoeuvring experts. Angelfish, *Pterophyllum scalare*, and butterflyfish, *Chaetodon* sp. have a short, diamond-shaped body with high dorsal and anal fins, in common with many more species from spatially complex environments. Coral reefs and freshwater systems with dense vegetation require precise manoeuvres at low speed. Short, deep bodies make very short turning circles possible.

Economic swimmers. Sunfish, *Mola mola*, opah, *Lampris guttatus*, and louvar, *Luvarus imperialis*, are among the most peculiar fish in the ocean. They look very different but have large body sizes in common. The sunfish reaches 4 m and 1500 kg, the opah may weigh up to 270 kg and the louvar is relatively small with a maximum length of 1.9 m, and weight of 140 kg (Wheeler, 1978). Little is known about the biology of these open ocean animals and virtually nothing about the mechanics of their locomotion. They all seem to swim slowly over large distances. The opah will use its wing-shaped pectorals predominantly and the louvar has a narrow caudal peduncle and an elegant high AR tail similar to those of the tunas. The sunfish has no proper tail but the dorsal and ventral fin together are an extremely high AR propeller. Sunfish swim very steadily,

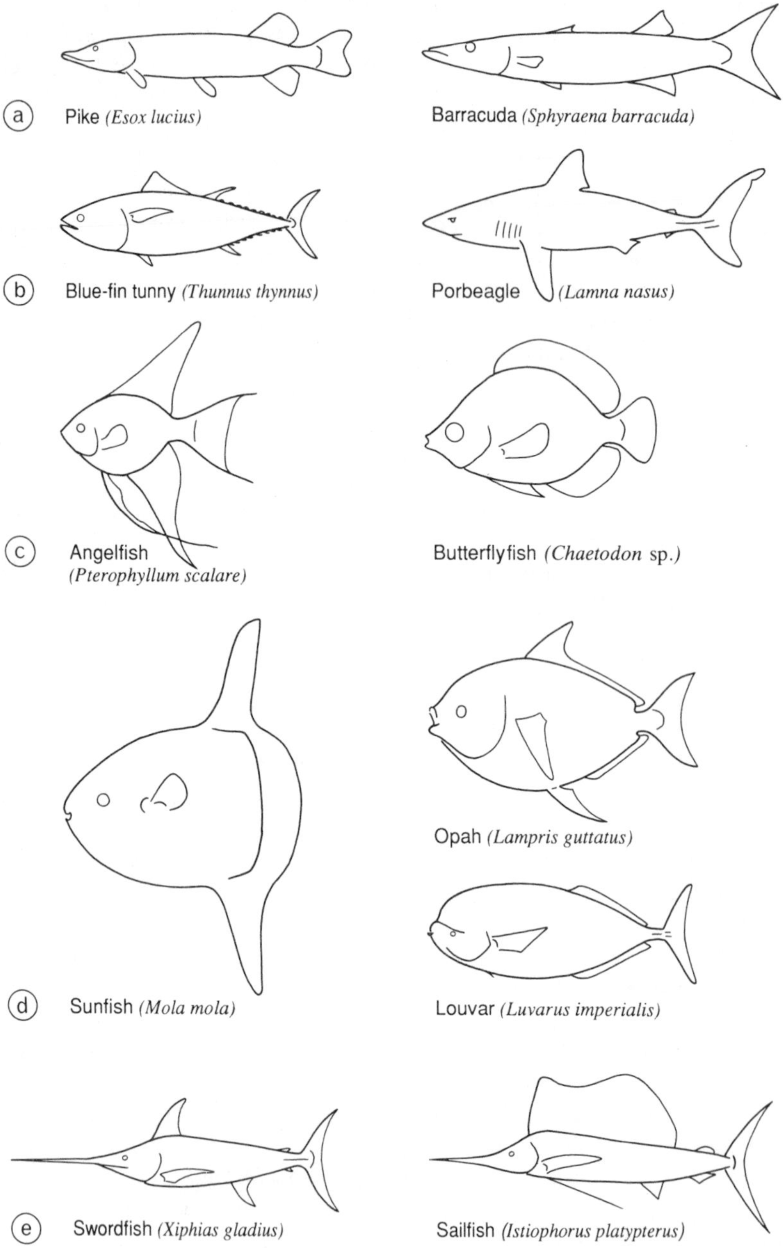
a
Pike (Esox lucius)
Barracuda (Sphyraena barracuda)
b
Blue-fin tunny (Thunnus thynnus)
Porbeagle (Lamna nasus)
c
Angelfish
(Pterophyllum scalare)
Butterflyfish (Chaetodon sp.)
d
Sunfish (Mola mola)
Opah (Lampris guttatus)
Louvar (Luvarus imperialis)
e
Swordfish (Xiphias gladius)
Sailfish (Istiophorus platypterus)

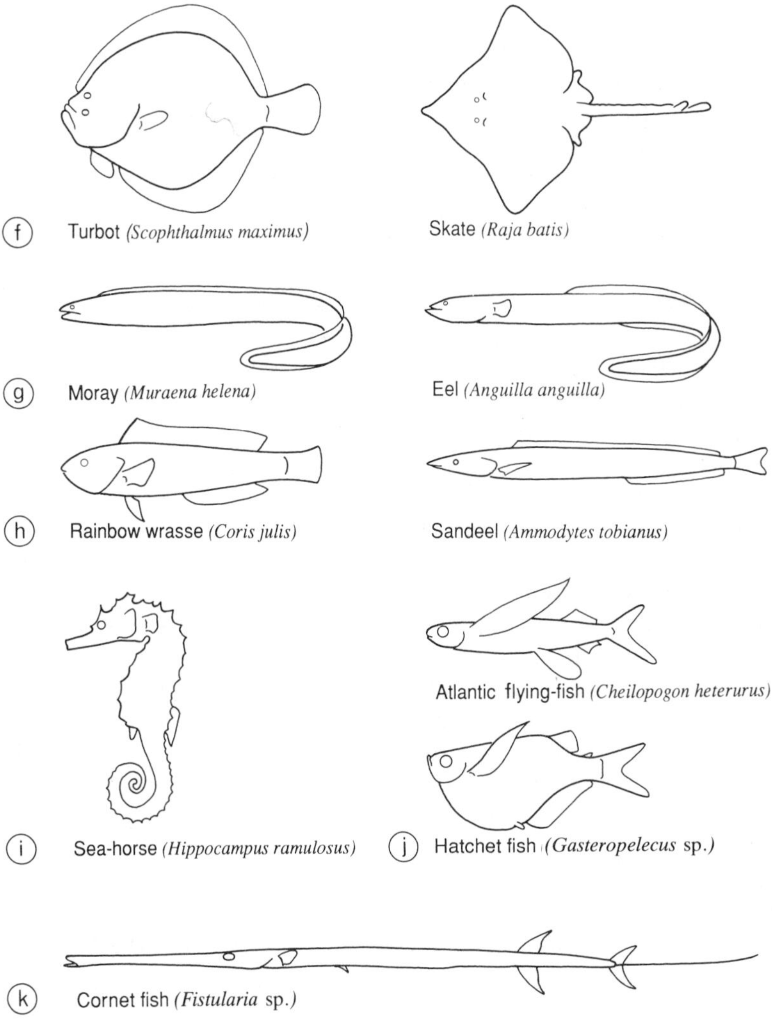

Fig. 4.7 Typical representatives of 11 swimming specializations: (a) accelerators; (b) stayers; (c) manoeuvrers; (d) economic swimmers; (e) high-speed specialists; (f) users of ground effect; (g) forward and backward wrigglers; (h) sandswimmers; (i) hoverers; (j) flyers; (k) recoil reducers.

moving the dorsal and ventral fin simultaneously to the left and, half a cycle later, to the right side (Damant, 1925). The dorsal and ventral fin have an aerodynamic profile on transection. The intrinsic muscles fill the main part of the body and insert on separate fin rays (Raven, 1939) enabling the sunfish to control the movements, camber and profile of its fins with great precision. Although there are no measurements to prove this as yet, my hypothesis is that these heavy species specialize in slow steady swimming at low cost. Inertia helps them to keep up a uniform speed, while their well-designed propulsive fins generate just enough thrust to balance the drag as efficiently as possible.

Fish adapted to reach the highest speeds. Swordfishes (Xiphiidae) and billfishes (Istiophoridae) show bodily features that no other fish has: the extensions of the upper jaws, the swords, and the shape of the head. They are probably able to swim briefly at speeds exceeding those of all other nektonic animals, reaching values of well over 100 $km\,h^{-1}$. A 3 m swordfish swimming at 30 $m\,s^{-1}$ reaches a *Re* approaching 10^8. The sword of swordfish is dorsoventrally flattened to form a long (up to 45% of the body length) blade with sharp edges. The billfish (including sailfish, *Istiophorus* sp., spearfish, *Tetrapturus* sp., and marlin, *Makaira* sp.) swords are pointed spikes, round on transection and shorter (between 14% and 30 % in adult fish, depending on species) than those of swordfish. All the swords have a rough surface, especially close to the point. The roughness decreases towards the head. One other unique bodily feature of the sword-bearing fishes is the concave head. At the base of the sword the thickness of the body increases rapidly with a hollow profile up to the point of greatest thickness of the body. This point lies much more forward than, for example, on the body of Scombridae. The dorsal and paired fins are positioned rearward of this point. Sword-bearing fishes are commercially important because they contain large quantities of highly appreciated muscle. The caudal peduncle is dorsoventrally flattened, fitted with keels on both sides. These features and the extremely high AR wing-shaped tail blades with rearward-curved leading edges are hallmarks of very fast swimmers.

There is considerable controversy about the function of the swords. A rather anthropomorphic school of thought relates swords to piercing and killing prey victims. A more physically orientated opinion dedicates the extended rostrum to the generation of microturbulence to achieve drag reduction. Swords certainly did not evolve to sink wooden boats. The arguments in favour of the drag-reducing function are stronger than the observations supporting the killing-device purpose. The rough surface on the sword and the concave shape of the head co-operate to reduce drag and to avoid large-scale turbulence.

During swimming at a *Re*, for the whole fish, of 10^8, the *Re* near the point of the sword will still be small and reach values where laminar-to-turbulent transition starts to occur. The slender blades and spikes hardly generate any pressure drag, but their roughness facilitates the transition of the flow in their boundary layer from laminar to microturbulent. Further down the sword, in the direction of the body, the roughness decreases. The boundary layer will remain in its microturbulent state owing to the increase of *Re* with distance along the sword. The microturbulent flow probably behaves similarly to a laminar flow under a *Re* regime where this would not otherwise be possible. The hypothesis is that the steady decline of the drag coefficient with increasing *Re* under laminar conditions continues up to values of *Re* near 10^8. Fig. 4.6 shows that the drag coefficient on a rigid streamlined body

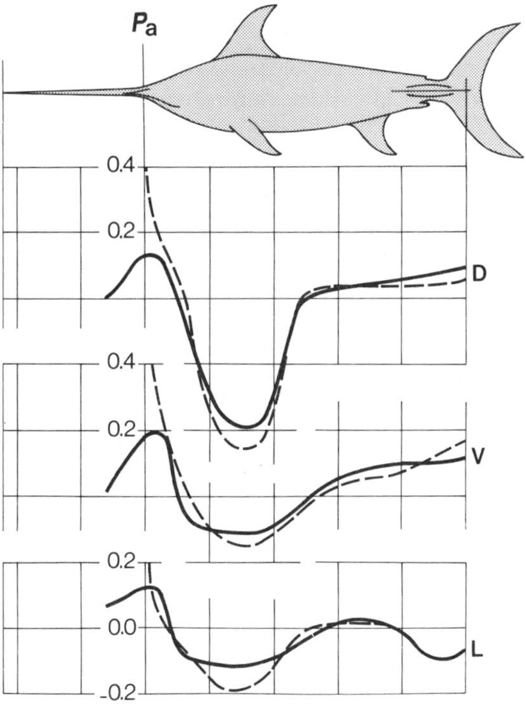

Fig. 4.8 Distribution of dynamic pressure coefficient p_a over the surface of a model of *Xiphias gladius*, along the dorsal (D), ventral (V) and lateral (L) midlines (solid curves). Broken curves are values for the same model but without the sword. Both models were tested at 6 m s^{-1}. The length of the complete model is 0.94 m, $Re = 6.7 \times 10^6$; the model without the sword was 0.65 m, $Re = 3.9 \times 10^5$. Modified from Aleyev (1977). ($p_a = 2(p - p_O)/\rho V^2$; p is the pressure (Pa) at the point of investigation, p_O is the static pressure (Pa) in the free stream, ρ is the water density (kg m^{-3}) and V is the velocity (m s^{-1}) of the free stream.)

could be approximately one fifth of the figure found for turbulent flow at *Re* 10^8.

The concave head at the base of the sword deflects the flow gradually, avoiding a sudden peak of dynamic pressure. Aleyev (1977) demonstrated this effect by comparing the pressure distribution along the dorsal, ventral and lateral sides of a wooden model of *Xiphias gladius* with that of the same model but without the sword (Fig. 4.8). Visualization of the flow in the boundary layer of a full scale model at the right speed could prove these hypotheses right or wrong.

Aleyev (1977) provides a survey of the relevant literature and was my main source of information.

Fish built to use ground effects. The shape of the body of flatfish and rays offers the opportunity to hide in the boundary layer close to the sea-bed. Arnold and Weihs (1978) used water tunnel experiments to demonstrate that only extremely fast flows are able to dislodge plaice, *Pleuronectes platessa*. There is another possible advantage connected with a flat body shape. Both flatfish and rays can be observed swimming close to the bottom. These fish are negatively buoyant and, like flying animals, must generate lift (a downwash in the flow) at the cost of induced drag to remain 'waterborne'. Swimming close to the ground could reduce the induced drag considerably, depending on the ratio between height off the ground and the span of the 'wings' (Norberg, 1990). This ground effect has not been investigated for flatfish as far as I know.

Experts in forward and backward swimming in mud and maze-type environments. Hunting moray eels (I followed the spotted moray, *Gymnothorax moringa*, hunting in daylight on a coral reef in the West Indies) use this technique, in a spectacular way, to search for prey in holes and crevices. Eels and congers can also quickly reverse the direction of the propulsive wave on the body and swim backward. Eels can move through thick mud and 'swim' through a stretch of grassland. The common feature of these fish is the extremely elongated flexible body. Swimming is usually not very fast and they prefer to swim close to the bottom.

Sandswimmers. Sandeels, *Ammodites tobianus*, and rainbow wrasses, *Coris julis*, sleep under a layer of sand, sandeels in daytime and rainbow wrasses during the night (Videler, 1988). Both species swim head-down into the sand using high-frequency low-amplitude oscillations of the tail. If the layer of sand is thick enough the speed is not noticeably reduced. Body shapes are similar, i.e. slender with a well-developed tail. Wrasses use their pectoral fins for routine swimming and move body and tail fin during escapes and to swim into the sand.

Hovering fish. Most neutrally buoyant fish species are capable of hovering in one spot in the water column. Some species can hardly do anything else. Sea-horses and pipefishes (Syngnathidae) rely on camouflage for protection from predators. They are capable of minute adjustments of the orientation of their body using high-frequency, low-amplitude movements of the pectoral and dorsal fins. (Sea-horses are the only fishes with a prehensile tail.)

Flying fish. Exceptionally large pectoral fins are used by the Exocoetina to make gliding flights out of the water, when chased by predators. Some species are four-winged because they use enlarged pelvic fins as well. We saw in Chapter 3 that the lower lobe of the caudal fin is elongated and remains beating the water during take-off. Hatchet fishes, Gasteropelecidae, actually beat their pectoral fins in powered flight. The pectoral fins have extremely large intrinsic muscles originating on a greatly expanded pectoral girdle. The flying gurnards, Dactylopteridae, have tremendously enlarged pectoral fins, usually strikingly coloured. There is still some dispute over their ability to use these large pectorals when airborne. Many more species occasionally or regularly leap out of the water (Lindsey, 1978) but are not adapted to such an extent as the real flyers mentioned here.

Recoil reducers. Cornet fishes (Fistulariidae) are predators of small fish in the littoral of tropical seas, most often seen above sea-grass beds or sandy patches between coral reefs. They seem to have two tails. The first one is formed by a dorsal and anal fin and the second one is the real tail. Beyond the tail there is a long thin caudal filament. I observed their hunting behaviour and saw that they are able to dash forward in one straight line without any side movements of the head, using large-amplitude strokes of the two tail fins and the trailing filament. My estimate is that the double tail fin configuration with the trailing filament serves to allow fast acceleration without recoil movements of the head. Precise kinematic measurements are needed to provide evidence for this assumption.

These special adaptations are only a few out of a wealth of possible examples. The interpretations are partly anecdotal and all highly speculative. I am not convinced that all other fish species are locomotor generalists. A closer study of the locomotory habits of a large number of species will show many more specialist groups than the odd eleven recognized here. We shall encounter several other possibilities (e.g. braking specialists) in Chapter 6 where swimming styles are compared.

4.5 SUMMARY AND CONCLUSIONS

The best pelagic swimmers among fish have streamlined bodies with thickness-to-length ratios between 0.22 and 0.24, which offers them minimum drag for the largest body volume. Dorsoventrally compressed parts of the body can easily be moved sideways. Laterally compressed parts counteract lateral movements, serve as pivot points for quick turns and have a stabilizing function during straight, forward swimming. A variety of structures enhances streamlining in fish.

The strongest layer of the skin is the stratum compactum, consisting of collagenous fibres arranged, similar to plywood, in layers with alternating fibre directions. The thickness of this mechanically important layer varies between species. The angles between the crossing fibres are usually between 50° and 70°, with a tendency to be larger just behind the head and reaching extremely small values at the caudal peduncle.

Published measurements of fish skin strength reveal large differences between species. Young's moduli vary tenfold from 0.4 GPa for shark skin to about 0.04 GPa for the skipjack tuna.

A broad overview is given of the variety of fish scale types. Possible locomotion-related functions are avoidance of transverse folds, conduction of the water flow and generation of microturbulence.

Mucus supposedly has a drag-reducing effect in fish, but evidence for this assumption is sparse. Measurements of the effect of mucus on the reduction of pressure drop in a flow pipe (the 'Toms effect') give highly variable results. The mucus of Californian bonito, for instance, has no drag-reducing effect, whereas the slime of the Pacific barracuda reduces the drag by 66%. The steepness of the gradient of the velocity profile is reduced by mucus in a rainbow trout, diminishing the viscous drag of the fish.

Most fish species are generalists as far as locomotion is concerned, and perform reasonably well in all aspects. Other species however are specialists in one or more facets of the swimming trade. Shape of the body and fins may indicate the direction of the specialization. Speculations about the special skills of 21 species, divided into 11 specialist groups, are given.

Chapter five

Fish kinematics: history and methods

5.1 INTRODUCTION

Kinematics is the study of movements without reference to forces. These studies concentrate on changes of position as a function of time.

The movements made by fish to propel themselves can be studied from two points of view. The first is a position on board the fish, where the motions of body and fins (especially amplitudes and frequencies) can be analysed relative to a fish-bound frame of reference. The second is an earth-bound position. From that standpoint the distances covered and the speeds, accelerations and decelerations achieved by the fish can be measured.

I start this chapter with a historical survey to get a feeling of how knowledge of fish swimming movements has developed. Subsequently, examples of modern studies will represent the present state of the art.

Kinematic analyses require fish to swim naturally under controlled conditions. Two fundamentally different approaches have been used to meet this requirement.

1. Fish have been induced to swim against a water current at various speeds.
2. Voluntary movements have been recorded of fish swimming in static water.

High-speed cine or video techniques together with the possibility to digitize certain salient points on the fish, or even the complete outlines of the fish images, frame by frame, are the invaluable tools for kinematic studies. Fish use predominantly oscillatory movements. These can be adequately described by the wave parameters: wavelength, period and amplitude. To avoid undue complexity, we will first concentrate on the wave parameters of the lateral body and tail movements during steady swimming of typical pelagic fish.

The explanation of the principles of the wave parameter analyses from film starts with a rather primitive method, requiring a pencil, a ruler and a pocket calculator. Subsequently, a more sophisticated computer-aided method will be discussed, where the wave phenomena are described by sine and cosine functions as series of Fourier coefficients.

5.2 HISTORICAL OVERVIEW

De Motu Animalium (On the Movements of Animals), written by Borelli, a professor in Naples, is my starting point for a survey of the development of our knowledge on fish kinematics, although occasional brief accounts by e.g. Aristotle and Pliny preceded this. It was published after his death in 1680. Borelli observed fishes (without indicating which species) and reached the conclusion that they do not use the pectoral fins for swimming. He thought that the pectorals were not strong enough and too flexible for propulsive actions; their functions were steering and braking. He tried to prove this point by cutting off the pectorals. The amputated fish managed to move quickly in all directions, which led him to conclude that the pectorals were not used for swimming. (In this case it is quite clear that one cannot expect to find a function of a structure by studying an animal where it has been removed. But even today, the first thought of many scientists, when asked to find the function of a structure, is to remove it.) Borelli found that the contortion of the tail caused the fish to swim: '*Instrumentum, quo pisces natant, est eorum cauda.*' He did not cut off the tail to prove this statement but drew Fig. 5.1(a) to explain the kinematics and the effect of the tail beat. According to his explanation, the folded tail blade is brought sideways to the most extreme lateral position (G). This displacement, he supposed, had no effect on the water because the webbed tail blade was contracted. During the subsequent quick active stroke, back to the straight position (C), the tail was extended maximally in dorsoventral direction, and pushed the water behind. Next is the passive stroke to leftmost (H) followed by a propulsive stroke back to the straight position. Note that Borelli took a point of view on board the fish to describe the movements. He interpreted the distortion of the shape of the body as a C-bend to the left and right. This could indicate that he studied the movements of fish out of the water, where this can actually be observed.

Borelli's views were not challenged for two centuries. Pettigrew (1873) found that during each sweep of lateral movement the tail is both extended dorsoventrally and flexed in the longitudinal direction of the fish, and concluded that there is no propulsive and recovery part of the stroke, but that the tail is an effective propeller during all parts of the stroke. His careful

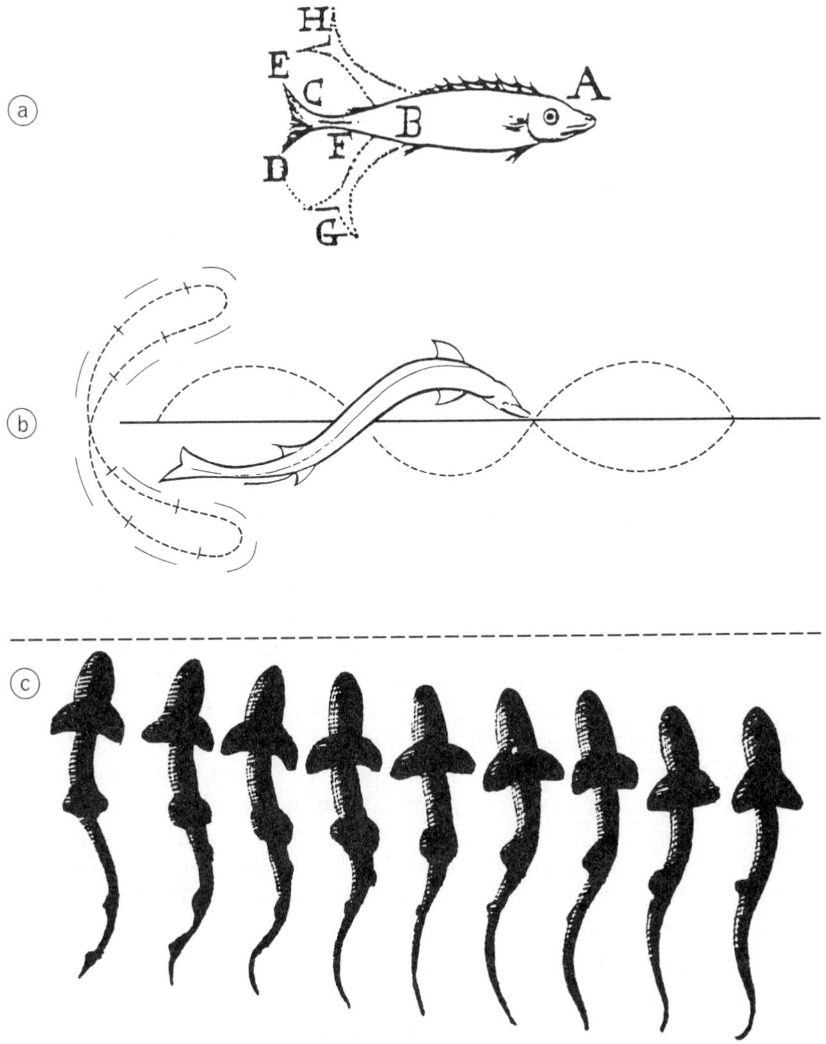

Fig. 5.1 Historic pictures marking the progress in early fish swimming studies. (a) Swimming movements observed by Borelli (1680). G and H are the most extreme lateral positions of the tail, C indicates the straight position of body and tail. The distance between D and E shows the dorsoventral extension of the tail. F is the flexible, and A–B the stiff part of the fish. (b) Pettigrew (1873) discovered the figure of eight described by the tail tip of the sturgeon. This figure can be observed when a trace is made of the path of the tail tip, while keeping the head in the same position along the line (solid curve) indicating the mean path of motion. (c) Houssay (1912) used Marey's motion pictures of a swimming dogfish. (a) and (b) are movements observed from a position on board the fish; (c) looks at the fish from the point of view of a fixed camera.

observations of swimming Atlantic salmon, *Salmo salar*, and sturgeon, *Acipenser sturio* (Fig. 3.1(b)), showed that there is not a single curve on the body, but a double curve (the body is S-shaped instead of C-shaped). Pettigrew also studied his fish from a position on the body, which enabled him to see that the lateral movements of the tail tip described a figure of eight during one swimming cycle.

Houssay (1912) published kinematic studies based on time series of photographs (Fig. 3.1(c)) of swimming dogfish, and eel. These motion pictures were taken and published in 1895 by Marey. The camera was mounted in a fixed position, so the frame of reference was earth-bound and no longer fish-bound. Houssay described how a lateral wave of curvature on the body can be observed moving from head to tail in elongated fishes. This wave caused the fish to advance through the water in the direction opposite to the direction of the wave.

Breder (1926) classified fish swimming movements into two main categories. He divided fishes into the ones that move body and tail and those that use the movements of appendages to propel themselves. In the first group he distinguished anguilliform, ostraciiform and carangiform swimmers and in the second group as many as seven styles were named after typical examples: amiiform, gymnotiform, balistiform, rajiform, tetraodontiform, labriform and diodontiform. Unfortunately, this widely used classification was not based on precise kinematic analysis and some of the styles are therefore rather ill-defined. That is why subsequent users modified the definitions. For example, the difference between anguilliform (after the eel, *Anguilla anguilla*) and carangiform (after the jacks or Carangidae) swimming is based on the position of a 'pivot point' and on the number of complete sine waves at each instant present on the body. Cinematographic analyses show that a pivot point or fulcrum (as between a boat and its rudder) cannot be found on fish like eel or jack. According to Breder, an anguilliform fish has more, and a carangiform swimmer less, than one-half of a sine wave on the body. Precise kinematic analysis (Videler and Hess, 1984) demonstrates that even a mackerel has more than half a wave on the body at any time. The crucial kinematic difference between mackerel and eel is the different increase of the amplitude of the lateral movement from head to tail; the amplitude increases almost linearly in the eel and follows a power function in the mackerel (Videler, 1981). The danger of Breder's classification is that it creates a false feeling of established knowledge which reduces the urge for further investigations.

Breder (1926) studied the functional significance of the tail of the rudd, by amputating it. The maximum escape velocity of this fish was found to equal that of an intact animal of the same size.

Precise studies of kinematics of swimming have been conducted by Gray (1933a) who started to describe the movements of eel, butterfish, *Pholis gunnellus*, whiting, *Merlangius merlangus*, dogfish, mackerel and greater sandeels, *Hyperoplus lanceolatus*, quantitatively. He found some important general rules. The waves on the body move faster backwards than the forward velocity of the fish. The amplitude of the lateral wave is different for different parts of the body and reaches a maximum near the tail tip. Each point of the body of a swimming fish follows an undulating path through the water. The wavelength of each path is less than that of the body of the animal. Gray also rediscovered the figure of eight at the tail tip and showed that in fact each point of the body travels in a horizontal figure of eight relative to the transverse axis, which is moving forward at the same average velocity as the whole fish. He showed that these movements are the mechanical results of the inextensibility of the body.

Gray's work generated considerable interest and investigations into animal locomotion, resulting in a large number of papers and books during the second half of this century.

Gray (1933b), among the rest, could not resist the temptation and removed the tail fin of a number of species to study its function. He found that the amputation did not substantially reduce the cruising speed of the fish but it altered the kinematics of the animals.

Propulsive waves of curvature

Numerous authors have described fish swimming movements, but Gray (1933a) showed most clearly how these can be understood as a combination of two wave-like phenomena:

1. cyclic changes of the curved shape of the body showing a lateral wave of curvature running in caudal direction at velocity v with a wavelength λ_b and a wave period T, where $v = \lambda_b T^{-1}$.
2. every single point of the body describes, in consequence of the wave of lateral curvature on the body, a sinusoidal track in a horizontal plane with forward velocity u, wavelength λ_s, period T and amplitude A, where $u = \lambda_s T^{-1}$.

Some simple models drawn in Fig. 5.2 clarify the relationships between the two wave systems. In these models a hypothetical fish is represented in a horizontal plane by a single line of a given length. Every line represents an instant of swimming motion with an equal time separation between adjacent lines. It is assumed that forward motion of the line is resisted by the water and that appropriate lateral movements generate force in interaction with the water. Forward movement is uniform and thrust equals drag. The direction of the body wave is to the right and swimming direction to the left. Fig 5.2((a) to (c)) illustrates the case where the amplitude A (the excursion in the Z-direction) is equal for all parts of the line. In Fig.

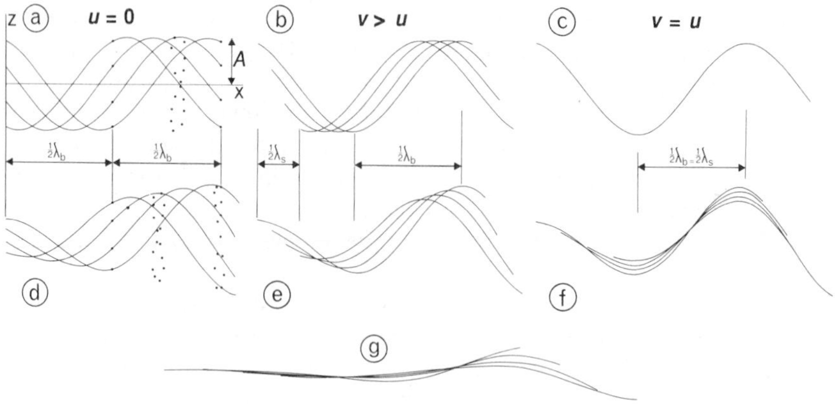

Fig. 5.2 Simple models to clarify the relationships between the two wave systems characterizing steady undulatory swimming in fish. The models are explained in the text. Reproduced with permission from Videler (1981).

5.2(a), forward speed u is 0; wave crests move to the right and show the velocity of the body wave v and the wavelength λ_b. The wave period T equals $\lambda_b v^{-1}$. The foremost point of the line oscillates along a straight line in the Z-direction with the same period T; λ_s and u are both equal to 0. Points of the body at distances $0.5\lambda_b$ and λ_b from the left side of Fig. 5.2(a) move also in a straight line of displacement in the Z-direction. All other points describe a figure of eight as indicated by a series of dots.

In Fig. 5.2(b), $u > 0$. The foremost point of the line proceeds to the left along a sinusoidal path with wavelength λ_s and speed u. The velocity of the displacement of the wave crests, v, to the right is greater than u. Fig. 5.2(c) shows the extreme case in which the forward speed u is equal to the speed of the body wave v. λ_s equals λ_b and therefore the line is following one single path.

Fish swim with an amplitude increasing along the body. Fig. 5.2((d) to (f)) is therefore slightly more realistic because these are drawn with a linear amplitude increase. Note in Fig. 5.2(d) how the foremost point and the point at $0.5\lambda_b$ from the left side of the figure follow a nearly straight track in the Z-direction. These two points will therefore proceed at a steady forward velocity in Fig. 5.2(e). The forward velocity of all other points in Fig. 5.2(e) will oscillate slightly during uniform motion, owing to the figure-of-eight effect. Fig. 5.2(f) shows that with increasing amplitude it is no longer possible for all points of the line to follow the same path when $v = u$.

Fig. 5.2(g) shows the movements of a 0.42 m cod swimming at a uniform velocity of 0.9 m s^{-1}. Movements of this fish have been recorded from the dorsal side. Each line is made to divide the silhouette of the fish into a right and a left half. Time separation between the lines is 0.05 s. The same parameters of the two-wave system can be used to describe the movements of the body of a swimming fish; this will be demonstrated in Section 5.4, but first I will deal with the methods used to study fish movements.

5.3 METHODS FOR KINEMATIC STUDIES ON FISH

Techniques to induce swimming under controlled conditions

Four fundamentally different techniques have been used to study swimming under experimental conditions.

1. Fish were trained to swim along a straight track in large tanks under static water conditions.
2. Swimming in large circles was studied in a 10 m diameter annular tank in the Marine Laboratory in Aberdeen.
3. Annular swimming chambers (fish wheels) were rotated against the swimming direction of the fish.
4. Fish were induced to swim against the water flow in stationary flume tanks.

The first method requires training of fish to swim back and forth between two feeding points by association of underwater flashing lights with the appearance of food (Wardle and Kanwisher, 1974). Usually fish will learn the trick very quickly and start to respond to the lights after a few trials. It has been used to study kinematics of cod (Wardle and Reid, 1977; Videler and Wardle, 1978; Videler, 1981) and saithe (Videler and Hess, 1984). Results obtained using this method with Atlantic salmon and the thin-lipped grey mullet, *Liza ramada* have not been published before and will be presented in Chapter 6.

In a slightly different approach, freshly caught mackerel were filmed while swimming fast along a straight path just after being released in a large tank (Videler and Hess, 1984; Wardle and He, 1988). The wave parameters of two species of sandeel (the lesser sandeel, *Ammodytes marinus* and the greater sandeel, *Hyperoplus lanceolatus*) have also been studied using this method: results for these species are presented as new data in Table 6.1.

The main advantage of methods where fish swim freely in static water is that fish are not hampered by limited space or interfering flow. It can be disadvantageous that fish are free to choose the speed they prefer, especially if a wide range of swimming speeds is required.

Using the second method, He and Wardle (1986, 1988) controlled the speed of mackerel, herring and saithe, using the optomotor reflex (the tendency of fish to keep station with a moving background or light pattern). A gantry across the radius of a circular tank moved around at angular velocities of 0–1 rad s^{-1}, which corresponded to 0–4.5 m s^{-1} at 9 m diameter where the fish were swimming in a 1 m wide annular channel. A slide projector mounted on the gantry cast a light pattern on the tank floor in the channel. It moved around at the speed of the gantry at 9 m from the centre. Fish followed this pattern and swam at the speed dictated by the moving gantry.

This method offers unconfined swimming space and static water conditions. At high velocities, kinematics are influenced by the effect of continuous turning (He and Wardle, 1988), but the effect is small. This method has proved to be excellent for studying endurance in fish (Chapter 10).

The principle of the fish wheel method was first described by Regnard (1893), improved by Fry and Hart (1948) and subsequently adapted by Bainbridge and Brown (1958). An annular channel, rectangular in cross section and filled with water, was rotated against the swimming direction at the speed of the fish. The fish remained stationary relative to the observer or camera. The animal chose its preferred swimming speed and swam in static water. Wall effects could be neglected as long as the fish swam near the centre of the channel and its size was small compared with the width of the channel. The largest fish wheel by Bainbridge and Brown (1958) with a diameter of 2.25 m was designed, using up- and down-sliding doors, to avoid slippage between water and the channel wall during fast accelerations to higher speeds. The rotational velocity of the wheel was proportional to the average swimming speed of the fish. The response time of the operator to the movements of the fish was potentially a serious source of error. It was expected to be small during steady swimming and to increase when swimming became more erratic. A further disadvantage of fish wheels was that weight problems limited their size, resulting in swimming along tight circular paths. Bainbridge (1958a, b) used the technique to study kinematics as well as performance.

Water tunnels or swimming flumes are especially popular among physiologists for metabolic studies at a range of speeds. Beamish (1978) describes the variety of the designs of these water treadmills. Fish are forced to swim against the flow between an upstream and a downstream wire mesh, the downstream one commonly being electrified. Closed tunnels are usually small because they are normally used as respirometers. Flumes need powerful pumps to create sufficient flow. The confined space, the noise and non-laminar flow regimes might be expected to cause anxiety to the fish and affect their swimming behaviour. Velocity corrections, based on theoretical considerations, are commonly used to compensate for the blocking effect of the fish on the flow in the tunnel and for wall effects. Webb's (1971a) corrections increase the velocity by 7.5% to 15%, depending on the size of the fish relative to the width of the tunnel. Flume experiments have provided a large amount of data about oxygen consumption and endurance, but not all the data reflect the best possible performance of the species involved. Webb (1993b) tested the effect of the width of a swimming channel on the frequency and amplitude of the swimming movements of trout. Fish in narrow flume tanks benefit from the close presence of walls especially at low speeds. Wall spacing had little effect on the maximum tail beat frequency, but maximum amplitude was reduced as wall width decreased. In other words, a fish can swim at the same speed with the same frequency using a smaller amplitude when the space between the walls is reduced. The effects were smaller at higher swimming speeds and porous walls had less effect on the kinematics than solid walls. However, the maximum prolonged speeds at which the fish could swim for 2 min in narrow channels were lower than those of control animals in a channel more than three times as wide. Narrow flumes obviously have positive as well as negative effects on swimming performance.

Kinematic studies of fish in flumes can be used to judge whether the swimming behaviour deviates substantially from optimal performance. For example, a fish is not considered to swim optimally if the distance covered per tail beat (the stride length) is on average much shorter than the maximum value found for the species concerned.

Cinematographic techniques

Cinematography and video are used to register the displacements of swimming fish and the movements of the body as a function of time. The displacements as a function of time can be used to calculate speed and accelerations, and the series of pictures provides frequencies and amplitudes of the movements of the body. High-speed cine cameras with intermittent film transport or video cameras with frame rates preferably of more than 100 frames per second are required to obtain sufficiently detailed time series of fast swimming movements of even the smallest fish with the highest tail beat frequencies. The camera should be in a fixed position, otherwise it is virtually impossible precisely to measure displacements with respect to an earth-bound frame of reference. Light is always a problem with high-speed photography of animals, because they tend not to behave properly under very bright light conditions. A sharp frozen image (with sufficient depth of field) of a fast movement demands short exposure times and a small aperture. Special tricks have been introduced by Wardle (1975) to overcome the light problem. His background illumination system was derived from film industry back-projection techniques. It was used to film trained fish in a 14 m long tank. The water was 0.8 m deep and the swimming channel was 1.2 m wide. A high-speed cine camera (Red Lake, Locam) was placed 2.8 m above the tank floor looking down. A transparent raft, floating on the water surface below the camera eliminated distortion of the images by waves. Two fibre-optic light sources were fitted vertically as close as possible around the lens. The beams from these point sources ran parallel to the optical axis of the lens. The light was almost fully reflected by high-reflex traffic sign material covering the bottom of the tank underneath the camera. The background, as seen by the camera, was very bright and each fish appeared as a dark silhouette with sharp outlines. The fish were not disturbed at all, because the overall light level in the tank was very low. The same light-reflecting system is applied on the road where a driver, sitting between the headlights in his car, clearly observes the reflecting traffic signs, because the optical axis of his eyes is the same as the direction of the reflecting light of the beams. From a position away from the car the reflection is not at all bright.

To enable subsequent accurate analysis, the camera has to be fitted with a timing device, offering the opportunity to measure the time between two successive images precisely. In video systems an accurate digital clock can be projected on every frame and the electronic system is inherently precise. In specialized cine cameras a light-emitting diode makes a series of time markers, usually every 0.01 s, on the film edge; this allows verification of the precision of the frame rate.

Film analysis and data processing

Unfortunately, despite training techniques as described above, fish hardly ever swim at a uniform speed along a straight line. This would make the analysis of the movements easy. Instead fish tend to make a variety of movements, where they may accelerate, decelerate, or quickly change direction. The active spreading and folding and beating of fins, which can be used in an endless variety of locomotory functions, makes selection of sequences rather difficult. The question one wants to answer must provide the selection criteria. For example, if the movements of body and tail at uniform swimming speeds are to be described, one does not want a fish braking by extending its pectorals during the sequence. This seems obvious but has frequently been overlooked. (See for instance Gray's (1933c) studies of whiting, where the distance covered per tail beat of the intact whiting is only one-quarter of the body length, which is a very poor performance for a species that can move ahead three-quarters of the body length per tail beat if it does not stick out its pectorals.)

A film of any movement, taken with a camera in a fixed position, running at a precisely known frame rate, makes it possible to measure displacements relative to a fixed, earth-bound, frame of reference in two dimensions. The velocity of the fish can be measured rather accurately from side views, only if the animal swims in a direction perpendicular to the optical axis of the camera. Slight deviations from this direction are not easy to measure. From such pictures it becomes directly obvious if a fish is not swimming at a constant horizontal level. The lateral shape of the body, and changes of that shape, can of course easily be obtained from side views.

Views from above or below contain more information because typical fish swimming movements are made laterally in the horizontal plane. It is difficult to detect slight changes in depth, but changes in direction are obvious. These views also reveal use of the pectoral fins. Knowledge of the velocity of the fish, the lateral amplitude of every part of the body from head to tail, the propagation of the propulsive wave, the frequency of the movements and the stride length or distance covered per tail beat, are obtained from the analyses of these pictures. In fact the track of every point on the body horizontally through the water can be reconstructed. Usually the length of the fish is used for the scaling of displacements, amplitudes, velocities and accelerations.

Films taken at high frame rates can be played at a lower rate to slow the motion. Although this usually gives an excellent impression of the elegance of the art of fish swimming, it does not provide the quantitative information

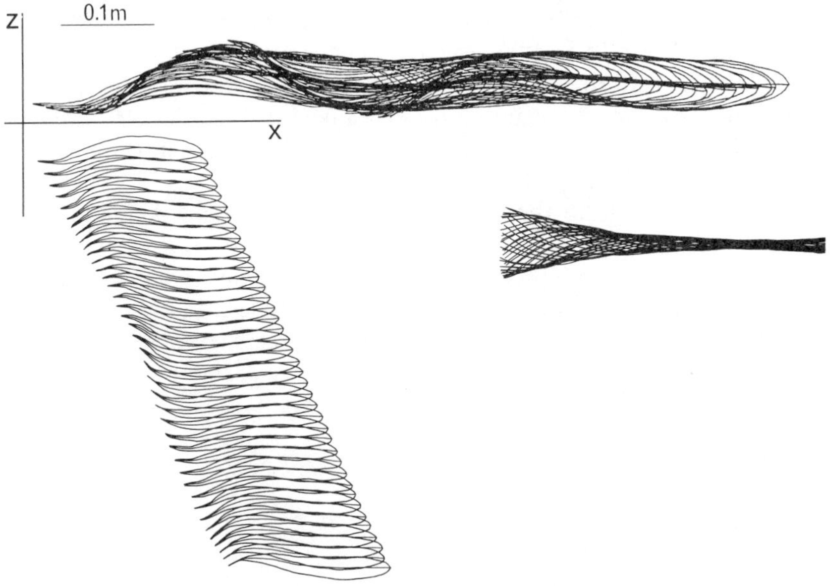

Fig. 5.3 Digitized outlines of images of saithe, *Pollachius virens*, in a fixed X–Z frame of reference, filmed at 100 frames per second. Top: superimposed outlines and computed centre lines. Left: the same images at half size shifted laterally for greater clarity. Right: centre lines only, with the X-positions of the nose made to coincide. Reproduced with permission from Videler and Hess (1984).

we want. Gray mounted the enlarged pictures of the films in a frame of reference where he shifted the successive frames laterally over a fixed distance. A background grid was used to keep the pictures in line. Nowadays we use computer-aided techniques to play around with digitized images of outlines of fishes on sequences of frames of film or video pictures which allow precise calculations of the kinematic parameters. Fig. 5.3 gives an example of the digitized outlines of saithe. A few examples of my kinematic studies are used to show how the main parameters can be determined.

5.4 EXAMPLES OF THE ANALYSIS OF STEADY, STRAIGHT SWIMMING

A simple method applied to swimming cod

The swimming motions of a fish like cod are graceful and the dynamic range of swimming performance is impressive. In its simplest form, waves of curvature run from head to tail down the body, alternately on the right- and left-hand side. The cyclic changes of the curved shape of the body form a running wave, characterized by a wavelength λ_b and a wave period T. The amplitude $A(x)$ is half the total lateral excursion of each point x of the body between head ($x = 0$) and tail ($x = L$). Its maximum value is usually found for $A(L)$ at the tail tip. The propulsive wave has a speed v of propagation down the body. An important derived parameter is the angle Θ, defined as the angle between the surface of the body and the mean path of motion in the horizontal plane.

Every single point on the body of the fish describes an undulating track through the water propagating at a mean speed u, which is the forward speed of the fish. Gray (1933a) found that the average wave length λ_s of these tracks is the same for all parts of the body and less than the average λ_b. The period of the wavy tracks of all the points of the body is the same as the period of the wave of curvature, T. λ_s is in fact the distance covered per tail beat or the stride length of the fish.

Fig. 5.4 (a) Lateral and dorsal view of cod 'plus', redrawn from film images. The positions of the points a–g on the dorsal side of the body could be identified on every frame. (b) Displacements in an X–Z frame of reference of the points a–g during straight forward swimming at a mean velocity of 0.9 $m\,s^{-1}$. For convenience, the x-axis has been shifted laterally for the track of each body point. The tracks can be recombined using the reference mark + in the left margin. (c) The time of passage of a wave crest through the positions a–g. The slopes of the regression lines through a series of points in this graph represent the speed of successive propulsive waves v_1, v_2 and v_3. Reproduced with permission from Videler and Wardle (1978).

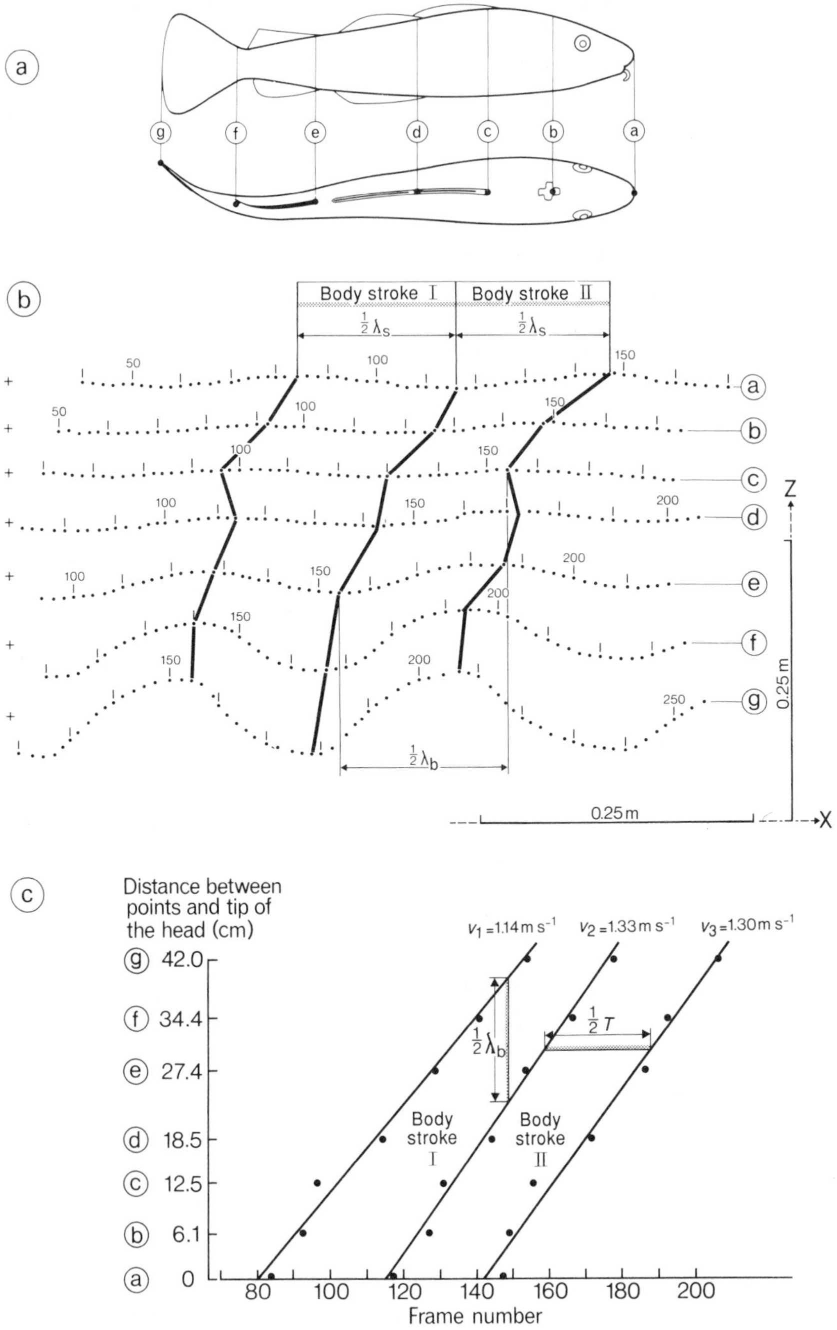
a
b
Body stroke I
Body stroke II
½ λs
Z
0.25 m
X
c
Distance between points and tip of the head (cm)
g 42.0
f 34.4
e 27.4
d 18.5
c 12.5
b 6.1
a 0
v1 = 1.14 m s-1
v2 = 1.33 m s-1
v3 = 1.30 m s-1
½ λb
½ T
Body stroke I
Body stroke II
Frame number

Measurements of these kinematic parameters can be achieved quite accurately as long as the wave crests along the body are well pronounced and the amplitude is substantial, even near the head. This is only the case in eel-like fishes. The analysis of one sequence of film images of the swimming movements of a 0.42 m cod is used as an example to show how these parameters can be found in a more fusiform fish, using a simple method (Videler and Wardle, 1978). The fish could be individually recognized because it had a '+' cold-branded on the top of its head. Sufficient light was used to see the dorsal fins and this brand. On the dorsal side of the body of cod 'plus', seven points (a–g in Fig. 5.4(a)), were identified on every frame. The time between the frames is 0.01 s. Fig. 5.4(b) shows the undulating tracks of the points a–g in an X–Z frame of reference. The mean path of motion of each track is parallel to the x-axis. The tracks represent the changing positions of the points a–g. The x-axis is, of course, the same for the tracks of all the points. To enable convenient study, the tracks in Fig. 5.4(b) are separated and labelled appropriately. The displacements in the X- and Z-direction are on the same scale. Those in the X-direction can be used to calculate the forward velocity of each of the seven points (the average velocity of every point of the fish is of course the same). The displacements in the Z-direction represent the amplitudes of the lateral movements. Instantaneous forward velocities (dx/dt) are the first time derivatives of the displacements. We used the five points differentiation formula of Lagrange to calculate dx/dt at a given time t = n:

$$dx/dt = f\{1/12x_{(n-2)} - 2/3x_{(n-1)} + 2/3x_{(n+1)} - 1/12x_{(n+2)}\} \quad (5.1)$$

where f is the number of frames per second. Speed u is the average of all the dx/dt values.

In theory, instantaneous acceleration or deceleration can be calculated by differentiating the dx/dt values with respect to time, using the same Lagrange equation. In practice, however, the dx/dt values are fairly erratic owing to imperfections of the measurements, and first need smoothing, for example by some running average method.

We drew three lines in Fig. 5.4(b), connecting the lateralmost positions to the left and right of each of the seven tracks. These lines illustrate how crests of the wave of curvature run down the body of cod 'plus' from head to tail. The horizontal distance between the outer lines is the wavelength λ_s in each track.

The parameters of the wave on the body, v, T and λ_b, are found by determining the time of passage of a wave crest through each of the seven points, starting from the head and following it down to the tail. In Fig. 5.4(c) these instants, expressed as film frame numbers, are plotted against the position of each point along the length of the body. The angles of the regression lines through these points represent the wave velocities v. The

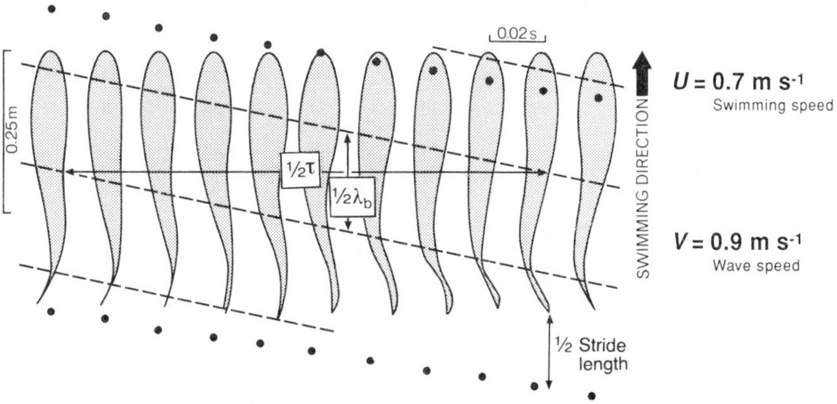

Fig. 5.5 Summary of the kinematic data of a 0.42 m long cod, swimming at a uniform speed of 0.7 $m\,s^{-1}$. The fish moves forward with respect to the background, represented by two black dots for each fish image (frame speed 50 frames s^{-1}). The images have been shifted to the right over a fixed distance, equivalent to a time shift of 0.02 s. Broken lines connect backward-moving crests of the propulsive wave on the body.

average horizontal distance between the outer lines gives a good estimate of period *T*, and the average vertical distance is the mean value for λ_b. Cod 'plus' was swimming at 0.9 $m\,s^{-1}$ with an average wave speed *v* of 1.27 $m\,s^{-1}$, a wave period *T* of 0.28 s, a stride length, λ_s of 0.25 m, and a wavelength of the body wave, λ_b of 0.37 m. The fish covered 0.6 body lengths per stride. The fish is progressing through the water while slipping, because its forward speed is less than the backward speed of the propulsive wave. The ratio of u/v is 0.73, which gives an indication of the amount of slippage. In a no-slip situation, u/v would be 1.

Fig. 5.5 summarizes the kinematic analysis of another sequence of steady swimming of cod 'plus' in a slightly different way. The images of the outlines are shifted laterally over a fixed distance representing 0.02 s time delay between the pictures. The fish swims forward relative to the background indicated by the black dots. The kinematic parameters of this sequence of pictures of the fish were calculated using the method described above.

As we saw in Chapter 1, the end of the tail blade determines to a great extent the direction of the flow of water that is left behind by the fish. The angle Θ and the velocities in the *X*- and *Z*-direction of the tailblade are therefore of great importance. Fig. 5.6 is a combined graph of these parameters against time of the 0.9 $m\,s^{-1}$ swimming sequence of cod 'plus' analysed in Fig. 5.4. It shows that dz/dt, the lateral velocity, and Θ reach the highest values in the beginning of the stroke, starting from each lateralmost position. When the tail blade is maximally bent in the middle of the

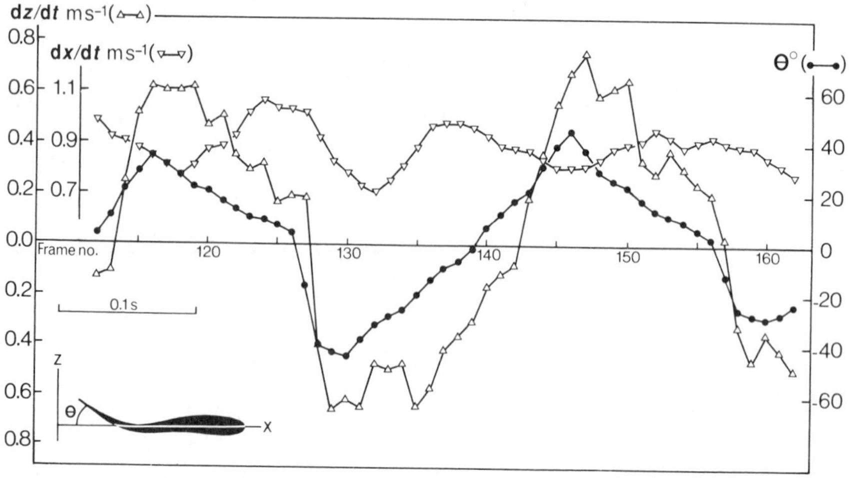

Fig. 5.6 Instantaneous velocities in the X and Z directions of the tail tip and instantaneous values of angle Θ between the end of the tail blade and the x-axis of a 0.42 m long cod. The fish is swimming at a steady velocity of 0.9 $m\,s^{-1}$. Reproduced with permission from Videler and Wardle (1978).

stroke, dx/dt is low. The oscillation of the forward velocity of the tail tip reflects the influence of the horizontal figure of eight first described by Pettigrew (1873). Fig. 5.6 represents the pattern commonly found for steadily swimming pelagic fish.

Computer-aided analysis of steadily swimming saithe

Swimming by travelling waves of lateral curvature with increasing amplitude is a periodic process. Therefore, the equations used in physics to describe harmonic motion can be applied to describe steady swimming motion of fish very accurately. Videler and Hess (1984) analysed fast continuous swimming of saithe and mackerel in this way. Outlines of top and side views of cine pictures of saithe (Fig. 5.7) illustrate the streamlined nature of the body. During steady swimming the pectoral and pelvic fins are pressed against the body and the anterior parts of dorsal and anal fins are folded in. The side view shows how the last dorsal and anal fins are partly erected and affect the shape of the fish during swimming. High-speed film sequences of top views of regular periodic swimming along a straight path were selected for the analyses. The outline of the image of the fish on each frame and two selected points of the reference grid were digitized and stored in the memory of a computer. Standard linear regression equations were applied to the coordinates of the tip of the head and the reference points to

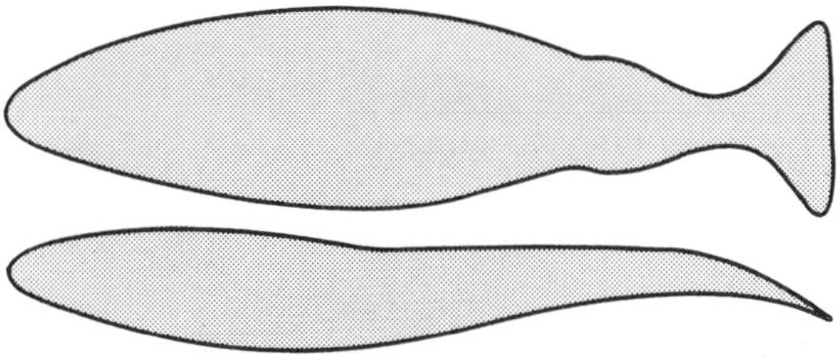

Fig. 5.7 Outlines of lateral (side) and dorsal (top) views of steadily swimming saithe, drawn from cine pictures. Reproduced with permission from Videler and Hess (1984).

calculate the mean path of motion. This path was designed to be the x-axis in a new frame of reference and all coordinates were transformed accordingly. The z-axis was perpendicular to the x-axis and horizontal. For each of the body outlines, a central line dividing the fish image into two lateral halves was computed. The example sequence is drawn in Fig. 5.3, showing the digitized outlines and computed centre lines of a steadily swimming, 0.35 m long saithe. The centre lines, consisting of 100 equidistant points, were used for the kinematic analysis of the swimming motion.

The following analysis offers a rather detailed account of a formal mathematical approach to determine the parameters of oscillating movements. The reader who is not interested in such details is advised to continue with the last paragraph before Section 5.5. This analysis deals with: the period T, the forward motion, the lateral displacements and the body curvature. The Z-position of each point on the centre line oscillates in time. The time intervals between successive extreme lateral positions are estimates of half the time period T. The resulting values for T are averaged, after giving each of the 100 points a weight proportional to the corresponding amplitude. T was 0.287 s in the example given and this value is used as a unit of time. The forward motion of the fish is approximated by a straight-line motion with a constant acceleration or deceleration plus a speed fluctuation originating from the tail beat. The mean distance between nose and tail, L, was used as a unit of length. (This is about 1% or 2% smaller than the length of the fish when it is straight.) The forward speed of the fish is calculated from the displacements of a point on the centre line close to the fish's centre of mass. Its X-position, x_m as a function of time is approximated by:

$$x_m(t) = a_0 + b_0(t-t_c) + c_0(t-t_c)^2 + a_1\cos 2\omega t + b_1\sin 2\omega t \quad (5.2)$$

where ω equals $2\pi/T$, and t_c is the time point half way between the first and last frame. The first three terms represent a motion with constant acceleration, and the two other terms a periodic fluctuation. A least-squares fit finds the best values for the factors a, b and c. The mean forward velocity u is

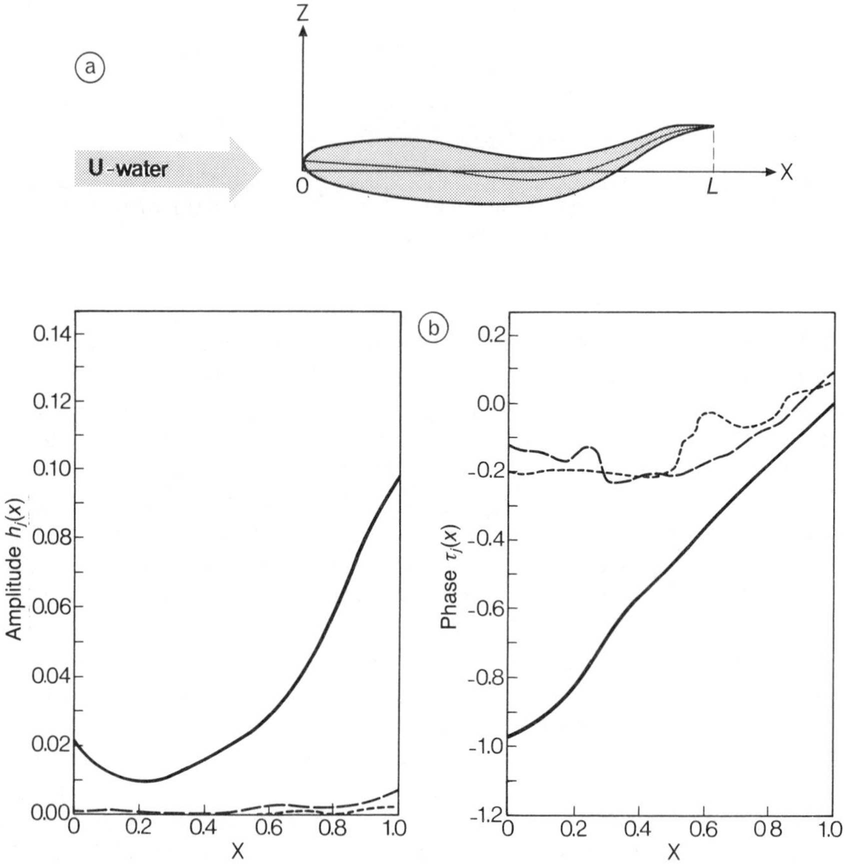

Fig. 5.8 (a) The coordinate system X, Z of a swimming saithe in dorsal view. The X-coordinate of the point of the head is 0 and that of the tail tip is 1, representing the length of the fish *L*. The water moves relative to the fish with swimming velocity *U* in the X-direction. This coordinate system is used to compute the lateral deflections in the Z-direction, *h*, of the calculated midline as a function of *X* (between $x = 0$ and $x = 1$) and time *t*. (b) Lateral deflections $h(x,t)$ for steadily swimming saithe, computed for the first three odd frequencies ($j = 1, 3, 5$) of a Fourier analysis. Left: the amplitude, $h_j(x)$, of the lateral movements expressed as fractions of the length *L*. Right: phase curves, $\tau_j(x)$, expressed as fractions of the period of the propulsive wave *T*. The abscissa of both graphs is the position on the fish between the point of the head ($x = 0$) and the tip of the tail ($x = 1$). The solid curves are the contributions of the coefficients of the sine and cosine functions of the first Fourier frequency. The broken curves are the contributions of the third, and the stippled curves those of the fifth frequency. Reproduced with permission from Videler and Hess (1984).

represented by the value found for b_0. For the comparison of swimming motions of fishes of different sizes and swimming at different speeds, it is very convenient to use dimensionless quantities. The forward velocity U, expressed dimensionlessly in lengths/period, is ($b_0 \times T/L$), which turned out to be 0.86 L/T (= 3 $L\,s^{-1}$ = 1.05 $m\,s^{-1}$) in our example, and the mean dimensionless acceleration is ($2c_0 \times T^2/L$) which was 0.02 L/T^2 (= 0.01 $m\,s^{-2}$). U is in fact the stride length, which is about 0.3 m in this case.

Now we will deal with the lateral displacements, assuming that the forward motion of all body points is uniform. This is convenient because the X-component of the motion can be ignored. The fish occupies a region between $x = 0$ (nose) and $x = L$ (tail) in the frame of reference drawn in Fig. 5.8(a).

The centre line of the fish is represented by the equation:

$$z = h(x,t) \tag{5.3}$$

We found that the lateral motion can be accurately described by six Fourier coefficients, a_j and b_j, using the equation:

$$h(x,t) \approx \sum_{j=1,3,5} [a_j(x)\cos(2j\pi t/T) + b_j(x)\sin(2j\pi t/T)] \tag{5.4}$$

where we only use the first three odd Fourier terms (1,3 and 5) because the movements are laterally symmetric. The contributions of the higher frequencies (7 and higher) are very small and drown in the noise. This last point becomes evident from Fig. 5.8(b), where the results obtained for the saithe of the sequence of Fig. 5.3, show that even the contributions of the third and the fifth frequencies are marginal. The left-hand graph gives the maximum amplitude as a function of the position on the body from head to tail. The graph on the right shows how the wave crest runs down the body from head to tail. The first frequency contribution uses almost exactly one period T to cover that distance. The higher frequencies only contribute slightly near the tail. The slope of the solid curve gives a good estimate of the body wave velocity v, which was 1.03 L/T (= 3.59 $L\,s^{-1}$ = 1.26 $m\,s^{-1}$). The slip ratio u/v was 0.84. This indicates that this saithe slips less than the cod we studied in the previous example, where we found a u/v of 0.73.

The first derivative of the lateral displacement $h(x,t)$ with respect to x is sin Θ, so we can use this method to obtain an approximate value of the angle between the fish and the mean path of motion as a function of t and x. It is approximate because Θ is calculated for the midline and we expect therefore the best results at the tail blade where the thickness of the fish can be neglected. The second derivative of the lateral displacement $h(x,t)$ with respect to x is a measure of the lateral body curvature. In fact this second derivative equals $1/r$, where r is the radius of curvature.

The 0.35 m saithe, shown in Fig. 5.3, is swimming steadily with simple harmonic motions at 3 $L\,s^{-1}$. The period of the cyclic lateral motions is 0.287 s. The amplitude increases from head to tail. The backward velocity of the propulsive wave was 3.59 $L\,s^{-1}$ and the distance covered per stride was 0.86 LT^{-1}. The forward velocity is 0.84 times the velocity of the propulsive wave. This kinematic approach forms the basis of a dynamic analysis of the interaction between fish and water, explained in Chapter 8.

5.5 SUMMARY AND CONCLUSIONS

The first section of this chapter gives a brief historical overview of studies of fish movements. Several attempts have been made to classify fish swimming styles. The application of cine film made precise studies of fast movements possible. Fish swimming with lateral undulations of body and tail have been studied extensively. These fish use a wave of curvature, running down the body with increasing amplitude, to propel themselves. Such movements can be described in terms of a two-wave system. A wave of curvature runs with increasing amplitude from head to tail with speed v, wavelength λ_b, and period T. Every point of the body moves forward following a sinusoidal track. The wavelength of these tracks is λ_s, the period is the same as the period of the wave of curvature and the speed is the swimming speed u. During steady swimming at uniform speed u is smaller than v.

Techniques used to induce swimming under controlled conditions are reviewed, and the advantages and disadvantages of each are discussed. Some important cinematographic aspects are addressed.

Two examples of kinematic studies of the steady swimming movements of a pelagic fish are given in detail. The first one follows the undulating movements of a small number of points on the fish and shows how instantaneous velocity is calculated from successive displacements. The second method is based on the movements of midlines of fish images calculated from digitized outlines on successive frames of film. Fourier analysis of the undulatory motions of 100 points on the midline provides precise values for the kinematic parameters required for the analyses of swimming dynamics.

Chapter six

Fish kinematics: swimming movements stride by stride

6.1 INTRODUCTION

Kinematic data are used to compare quantitatively the different swimming styles representing adaptations to the different niches within the aquatic environment. We will find common features and specific differences.

The examples in the previous chapter focused on harmonic motions during steady straight swimming at uniform speed. Kinematic data collected for this type of swimming are used to define general features of swimming with undulations of body and tail and to compare swimming styles.

Among fish in the wild, however, steady swimming is exceptional. Acceleration and deceleration during burst-and-coast swimming, fast starts and quick turns are common during natural swimming behaviour. Tilting, a peculiar behaviour shown during low-speed swimming, is described and discussed. We will survey the few explorations in this complex area of fish kinematics.

The last two sections of this chapter focus attention on the kinematics of fins. One treats the subtle movements of the tail blade, and the next is on swimming with other appendages, with emphasis on the use of pectoral fins.

6.2 COMPARISON OF KINEMATIC DATA

Table 6.1 compiles accurately determined propulsive wave characteristics for fish swimming at steady speeds without the use of pectoral fins. The number of studies that meet these requirements and provide sufficient information is

Table 6.1 Propulsive wave characteristics for fish swimming at steady speeds using undulations of body and tail. For explanation of variables, see text

Species	*Length*	A(L)		*Freq.*	T	λ_b		λ_s		u		v		u/v	*Source**
	(cm)	*(cm)*	(L)	*(Hz)*	*(s)*	*(cm)*	(L)	*(cm)*	(L)	*(cm* s^{-1}*)*	(L s^{-1})	*(cm* s^{-1}*)*	(L s^{-1})		
Abramis brama	19.0	2.2	0.12	3.8	0.26	15.6	0.82	11.9	**0.63**	45.0	2.4	59.0	3.1	**0.76**	1
Leuciscus leuciscus	25.0	3.4	0.13	2.8	0.36	20.5	0.82	15.1	**0.60**	42.0	1.7	57.0	2.3	**0.74**	1
Carassius auratus	16.0	1.9	0.12	4.0	0.25	16.3	1.02	11.4	**0.71**	46.0	2.9	65.0	4.1	**0.71**	1
Esox sp.	18.3	1.0	0.06	6.0	0.17	15.0	0.82	7.6	**0.41**	45.8	2.5	90.9	5.0	**0.50**	2
Oncorhynchus mykiss	20.1	1.6	0.08	4.0	0.25	17.9	0.89	12.7	**0.63**	50.3	2.5	71.6	3.6	**0.70**	2
Oncorhynchus mykiss	5.5	0.6	0.10	9.9	0.10	5.3	0.96	3.6	**0.65**	35.5	6.5	53.0	9.6	**0.67**	3
	11.6	1.0	0.09	7.1	0.14	10.6	0.91	8.1	**0.70**	58.0	5.0	75.7	6.5	**0.77**	3
	24.9	2.5	0.10	3.6	0.28	20.5	0.82	15.9	**0.64**	57.0	2.3	73.2	2.9	**0.78**	3
	43.3	3.0	0.07	2.4	0.42	38.6	0.89	27.7	**0.64**	66.5	1.5	91.9	2.1	**0.72**	3
Salmo salar	66.5	6.0	0.09	1.3	0.76	62.9	0.95	39.0	**0.59**	51.2	0.8	82.7	1.2	**0.62**	7
	66.5	5.0	0.08	1.7	0.57	68.8	1.03	43.8	**0.66**	76.8	1.1	121.0	1.8	**0.63**	7
	66.5	3.0	0.05	5.3	0.19	61.0	0.92	45.6	**0.68**	237.0	3.6	322.0	4.8	**0.74**	7
Liza ramada	36.0	3.0	0.08	4.6	0.22	34.2	0.95	26.0	**0.72**	120.0	3.3	158.0	4.4	**0.76**	7
Gadus morhua	42.0	3.2	0.08	2.8	0.36	35.0	0.83	26.0	**0.62**	73.0	1.7	96.0	2.3	**0.76**	4
	42.0	2.9	0.07	3.3	0.30	34.0	0.81	27.0	**0.64**	91.0	2.2	114.0	2.7	**0.80**	4
	42.0	3.5	0.08	3.6	0.28	35.6	0.85	25.8	**0.61**	92.0	2.2	127.0	3.0	**0.72**	4
	69.5	5.9	0.09	1.9	0.52	63.4	0.91	44.8	**0.64**	86.9	1.2	123.0	1.8	**0.71**	7
Pollachius virens	35.0	2.3	0.07	2.2	0.45	32.2	0.92	28.1	**0.80**	63.0	1.8	72.2	2.1	**0.87**	5
	35.0	3.1	0.09	2.7	0.37	35.4	1.01	27.5	**0.79**	73.5	2.1	94.5	2.7	**0.78**	5
	35.0	3.2	0.09	3.3	0.31	36.8	1.05	28.8	**0.82**	94.5	2.7	120.5	3.4	**0.78**	5
	35.0	3.3	0.10	3.7	0.27	35.7	1.02	27.6	**0.79**	101.5	2.9	131.3	3.8	**0.77**	5
	35.0	3.4	0.10	3.5	0.29	36.1	1.03	30.1	**0.86**	105.0	3.0	125.6	3.6	**0.84**	5

	35.0	3.4	0.10	3.8	0.26	35.0	1.00	28.3	**0.81**	108.5	3.1	134.1	3.8	**0.81**	5
	35.0	3.5	0.10	5.3	0.19	34.3	0.98	23.0	**0.66**	122.5	3.5	182.4	5.2	**0.67**	5
	40.0	3.6	0.09	4.5	0.22	36.8	0.92	29.3	**0.73**	132.0	3.3	165.8	4.1	**0.80**	5
	40.0	3.1	0.08	4.9	0.21	36.4	0.91	28.0	**0.70**	136.0	3.4	176.7	4.4	**0.77**	5
Scomber scombrus	30.0	3.1	0.10	5.6	0.18	34.5	1.15	24.2	**0.81**	135.0	4.5	192.7	6.4	**0.70**	5
	31.0	3.5	0.11	5.5	0.18	31.3	1.01	27.5	**0.89**	151.9	4.9	173.0	5.6	**0.88**	5
	31.0	3.5	0.11	5.0	0.20	34.1	1.10	31.8	**1.03**	158.1	5.1	169.7	5.5	**0.93**	5
	31.0	3.6	0.12	12.0	0.08	29.5	0.95	22.1	**0.71**	266.6	8.6	354.8	11.4	**0.75**	5
	33.0	3.4	0.10	5.1	0.20	33.3	1.01	25.9	**0.78**	132.0	4.0	170.1	5.2	**0.78**	5
	33.0	3.6	0.11	5.4	0.19	35.6	1.08	30.7	**0.93**	165.0	5.0	191.6	5.8	**0.86**	5
	34.0	3.6	0.11	12.2	0.08	32.0	0.94	25.6	**0.75**	312.8	9.2	389.8	11.5	**0.80**	5
	34.0	3.6	0.11	13.9	0.07	34.0	1.00	25.0	**0.73**	346.8	10.2	472.2	13.9	**0.73**	5
	34.0	3.5	0.10	14.9	0.07	32.0	0.94	25.5	**0.75**	380.8	11.2	477.0	14.0	**0.80**	5
Anguilla anguilla	14.0	1.4	0.10	3.6	0.28	11.1	0.79	7.7	**0.55**	27.9	2.0	40.1	2.9	**0.70**	6
Hyperoplus lanceolatus	28.7	2.2	0.08	3.6	0.28	21.5	0.75	14.6	**0.51**	52.0	1.8	76.8	2.7	**0.68**	7
	28.7	2.0	0.07	3.0	0.33	21.2	0.74	15.4	**0.54**	46.5	1.6	64.2	2.2	**0.72**	7
	28.7	2.6	0.09	2.7	0.37	22.7	0.79	14.5	**0.51**	39.2	1.4	61.4	2.1	**0.64**	7
	28.7	1.7	0.06	2.9	0.34	22.7	0.79	14.9	**0.52**	43.8	1.5	66.8	2.3	**0.65**	7
	33.0	3.6	0.11	2.6	0.39	28.7	0.87	18.3	**0.56**	46.9	1.4	73.6	2.2	**0.64**	7
Ammodytes marinus	8.7	0.8	0.09	6.0	0.17	6.9	0.79	4.7	**0.55**	27.9	3.2	40.9	4.7	**0.68**	7
	8.2	0.8	0.09	7.7	0.13	6.6	0.81	4.8	**0.59**	37.2	4.5	50.8	6.2	**0.73**	7
	10.5	1.2	0.11	5.0	0.20	9.7	0.92	6.7	**0.64**	33.5	3.2	48.5	4.6	**0.69**	7
	8.7	0.9	0.10	3.5	0.29	6.9	0.79	3.1	**0.36**	10.7	1.2	23.8	2.7	**0.45**	7

**Sources: 1, Bainbridge (1963); 2, Webb (1988); 3, Webb et al. (1984); 4, Videler and Wardle (1978); 5, Videler and Hess (1984); 6, Hess (1983); 7, New data.*

small, probably owing to the elaborate filming and analytical techniques needed.

General trends

The smallest amplitude of the lateral movements is usually made by a point just behind the head. From there the amplitude increases somewhat towards the head and there is a substantial increase in the direction of the tail. The largest lateral displacements are made by the tail tip. The relative tail tip amplitude varies between 0.14 and 0.05 L. This variation seems larger than in fact it is, because the average value of 0.09 L has a standard deviation of just 0.02 L. Thus the maximum amplitude at the tail tip approximates 10% of the body length for most fish.

A strongly significant relation exists between the frequency (F in Hz) and the speed relative to the body length (u in $L\,s^{-1}$). The equation:

$$u = 0.71\ F \qquad (r^2 = 0.93;\ N = 45) \tag{6.1}$$

explains 93% of the total variation. This equation provides a fairly accurate first estimate of the speed of any fish, given its length and tail beat frequency. The average stride length predicted by Equation 6.1 is 0.71 L. This value is slightly higher than the average value for λ_s (L) of 0.68 L (SD 0.13 L) in Table 6.1. The frequency and the velocity can be more accurately determined than λ_s, therefore 0.72 L is the best estimate for the average stride length available.

There is also a significant relation between λ_s (L) and the u/v ratio:

$$u/v = 0.34 + 0.57\ \lambda_s\ (r^2 = 0.72;\ N = 45) \tag{6.2}$$

indicating that longer strides give higher u/v values and hence less slip.

The mean relative wavelength λ_b is 0.92 L (SD 0.10 L), so there is usually just about one complete wavelength on the body of a steadily swimming fish. This mean figure is, however, not all that important. Differences among species are more interesting, because these may provide insight into the relations between swimming style, body shape and ecological niche.

Specific differences

The bream, *Abramis brama*, the dace, *Leuciscus leuciscus* and the goldfish, *Carassius auratus*, of the first data set of Table 6.1, swam steadily in a fish wheel and were filmed from above at 50 frames s^{-1}. Bainbridge (1963) examined the reconstructed paths of the transverse movements of six points on the body of each fish to determine the kinematic parameters. The amplitudes of these three species are large in comparison with the rest of the data. Bainbridge studied the deformations of the tail blade closely and took the lateral excursions of the dorsal and ventral tips of the tails to measure the amplitude. This method gives slightly larger values than when the tail is treated as a flat plate extension of the midline of the body. This implies that the other amplitude data are slight underestimates.

Webb (1988) compared steady swimming kinematics of a predator and its prey. He studied the tiger musky, a hybrid pike: male *Esox lucius* × female *Esox masquinongy*, and the rainbow trout in a flume tank at a range of swimming speeds between about 1 and 4 $L\,s^{-1}$. Table 6.1 compares animals of approximately the same length, swimming at the same relative speed. The tiger musky swims with a smaller tail amplitude and a higher frequency. The length of the wave on the body is similar but the difference in frequency makes its velocity, v, higher in the pike. The stride length, λ_s, of the predator is surprisingly small, and the amount of slip, indicated by the factor u/v, is considerably higher than that of the potential prey. The rainbow trout is showing the signs of being a better constant-speed swimmer than the pike. The tiger musky is an ambush predator and is likely to be a more skilful fast starter.

The effect of size and speed on the steady swimming performance of the rainbow trout was studied in a large flume tank by Webb *et al.* (1984). From their results I selected for each of the four size classes the swimming sequences with the highest stride lengths. The kinematic parameters, $A(L)$, λ_b, λ_s, u, v and u/v in this selection are similar to those of the trout in the previous study if we compare the relative figures. Webb *et al.* found that the relative stride lengths were inversely related to body size and positively correlated with speed; their results implied that u/v also increased with speed. The speed relations in these results are not supported by our studies on saithe (Videler and Hess, 1984). The u/v ratio of 35 cm saithe does not rise as speed increases from 1.8 to 3.5 $L\,s^{-1}$; the amount of slip remains constant. Also the stride length of saithe is not in any way correlated with speed changes. I will come back to this controversy in the next section which is devoted to velocity and length effects on stride length.

The Atlantic salmon, the mullet and the large cod were raced between feeding points and studied following the method used for saithe. The kinematic data for the salmonids are very similar, although λ_b of the salmon seems to be slightly larger. One sequence of mullet has a higher length-specific λ_s than the salmonids but is similar in other aspects.

The 42 cm cod in Table 6.1 is cod 'plus' from Videler and Wardle (1978). The analytical technique used has been explained in detail (p. 104). There is no apparent size effect. The average stride length for cod is 0.63 L. The gadoids cod and saithe use different swimming strokes. Cod has a shorter stride length and a shorter relative wavelength λ_b than saithe. Saithe and mackerel are very similar reflecting the similarity of their life styles as pelagic predators. Cod is a demersal fish; its bottom-dwelling skills are more directed at manoeuvring.

The methods used to measure the kinematic data for saithe and mackerel have been explained in Chapter 5. It is important to note that in Table 6.1, sequences showing even the slightest deceleration have been omitted from the data set published by Videler and Hess (1984). The values for both stride length and u/v are high. In one case a mackerel managed a stride length exceeding its body length, slipping very little with a u/v of 0.93. Frequencies increase with speed, and the amplitudes and λ_b values remain fairly constant. The differences in fish lengths are too small to allow any conclusions on size effects.

Hess (1983) used the same method as the one for saithe and mackerel to obtain the eel data in Table 6.1. λ_b is 0.79 L which means that there is more than one wavelength on the body of a steadily swimming eel at any time.

The data for both species of sandeel were obtained with the procedure described for mackerel. The selected sequences represent steady swimming, without obvious accelerations and with no sign of the slightest deceleration. The parameters shown for the greater sandeel and the eel are very similar. The λ_b of the lesser sandeel is slightly larger. There is one important difference between sandeels and eels regarding the division of amplitude of the lateral movements from head to tail. The amplitude increases fairly linearly towards the tail in the eel but shows a power function, similar to that found for mackerel and saithe in both sandeels. In this context, it is interesting to recall Batty's (1984) measurements of the amplitudes of herring larvae. He found a linear increase of amplitude along the body of a 11 mm larva and a power function for a larva twice that size.

Batty (1981, 1984) also measured v and u for plaice and herring larvae. In fish larvae, viscosity dominates the swimming conditions to a large extent and these larvae swam with extended pectorals. These facts probably explain why the u/v values found by Batty were only between 0.2 and 0.4, indicating high slippage.

Stride length, size, frequency and speed

There are three key papers on the scaling of stride length with body size and speed. Bainbridge (1958a) measured size effects on steady swimming speeds and tail beat frequencies of trout, goldfish and dace in his fish wheel. Hunter and Zweifel (1971) studied the relationship between tail beat frequency and speed for jack mackerel, *Trachurus symmetricus*, varying in length between 4.5 and 27 cm. Webb *et al.* (1984) collected the same type of data for rainbow trout. Jack mackerel and trout were swum in different types of water treadmills. The results of these studies are virtually the same but the interpretation is not straight forward. I will use Webb *et al.*'s data to illustrate the results.

The swimming speed ($\mathrm{m\,s^{-1}}$) increases rectilinearly with the tail beat frequency (Hz). The slopes of the lines are different for each body length. The maximum tail beat frequency measured is also size related, showing the highest values for the smaller sizes. This results in a graph similar to Fig. 6.1(a) for each species, where a separate linear relation between swimming speed and tail beat frequency is shown for each size class. If the data of the

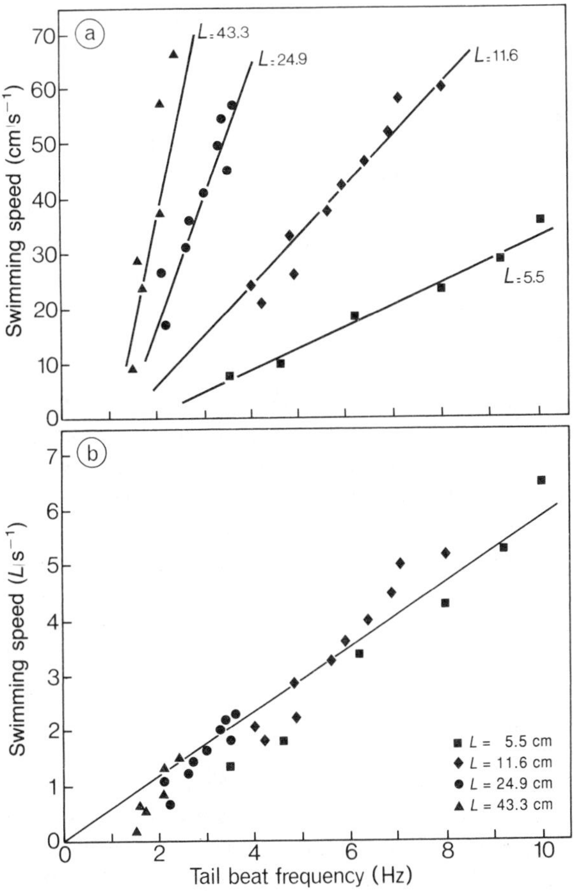

Fig. 6.1 (a) Relationship between swimming speed and tail beat frequency for four rainbow trout of different length, swimming in a flume tank. (b) Combined results expressing the speed in body lengths per second. The curve is the linear regression with 0 intercept calculated through the data points after discarding the three lowest frequencies for each size; stride = 0.59 *L*. Based on data by Webb *et al.* (1984).

different sizes are combined by expressing the speed in $L\,s^{-1}$, the linear relation is spoiled by data points near the lowest frequencies as demonstrated in Fig. 6.1(b). Bainbridge (1958a) tried to overcome this problem and created one rectilinear relation between relative speed and tail beat frequency by disregarding frequencies of less than 5 Hz. He did that despite the fact that small fish use higher frequencies than large ones. This excluded virtually all the data points for the larger individuals (e.g. 20 out of 22 data points for the largest, 22.5 cm, goldfish and only 4 out of 22 for the 7 cm fish, were excluded). At high speeds, forces to control yawing, pitching and rolling are likely to be small compared with thrust. The necessary stabilizing forces are probably generated by the caudal fin, involving numerous fine control movements such as those described by Bainbridge (1963). However, at low speeds the forces needed to control steering, buoyancy and equilibrium become relatively large compared with thrust, so fish erect unpaired fins and use their pectorals. The increased lateral surface and the moving pectorals will cause extra drag and hence result in a smaller distance covered per tail beat. The amplitude is often modulated at low speeds. All these considerations provide an excuse to exclude the slow swimming bouts from the analysis of the relation between relative speed and frequency. However, the limit below which data are excluded must be the minimum stable swimming speed, not a fixed frequency. These slow speeds and additional stabilizing motions occur at lower frequencies in larger fish.

Hunter and Zweifel (1971) corrected their data set by excluding low velocities. Above these velocities, fish usually swim steadily with depressed dorsal and anal fins and with pectorals in a fixed position, normally flat against the body. The amplitude is a constant fraction of the body length, and the ratio of speed ($L\,s^{-1}$) over frequency (Hz) has a fixed value which can also be expressed as a fraction of the body length. For Webb *et al.*'s trout in Fig. 6.1(b) this is approximately 0.59 *L*. This slope of the line in Fig. 6.1(b) is the average stride length for swimming bouts after data points of the three lowest tail beat frequencies in each size class were removed. This average stride length seems to be speed and length independent within the normal ranges of adult sizes and speeds, but it depends on the body shape and swimming style of a fish and therefore varies among species.

The maximum stride length that a fish species can achieve represents its top gear, which is independent of size and speed. Published data show that fish do not always use their maximum stride length during steady-speed swimming. Webb's (1986) lake sturgeon, *Acipenser fulvescens*, for example, can reach values up to 0.64 *L* per stride, but often use a lower gear, in some cases even as low as 0.25 *L*.

This size independence however does not include larval stages. Batty's (1981, 1984) plaice and herring larvae had stride lengths in the order of 0.3 *L* during steady swimming. My students found strides between 0.1 and

0.3 *L* for larval guppies, and of about 0.4 *L* for larvae of garfish, *Belone belone*, and rainbow trout. These larvae are swimming at low *Re* (Chapter 1) where viscosity dominates the events.

The maximum tail beat frequency of fish, and hence the maximum speed, are size and temperature dependent, as will be shown in Chapter 7.

6.3 UNSTEADY SWIMMING

Swimming at uniform velocities along a straight path is rather exceptional among fish. Unsteady movements are common in the behavioural repertoire of most species. The kinematics of fast starts, rapid turns, braking and burst-and-coast swimming are therefore of great biological relevance.

Unsteadiness has to do with rates of change of speed, i.e. accelerations or decelerations. These can be either directly measured using a calibrated accelerometer mounted on the fish, or calculated by double differentiation of displacements as a function of time. The displacements have to be accurately measured from high-speed film or video frames. Both procedures can introduce large mistakes. The accelerometers must be exactly at the centre of mass, or centripetal acceleration can generate extremely large errors (P.W. Webb, pers. comm. 1992). The order of magnitude of the mistakes owing to double differentiation depends on the film or video frame rate and the magnification used (Harper and Blake, 1989).

Fast starts

High accelerations during fast starts, as the first part of attack or escape behaviour, are biologically important for obvious reasons. We want to know how high these accelerations are and how they are achieved by the movements of the starting fish. Is the fast start performance of a predator better than that of its potential prey? Early studies to answer these questions were mainly descriptive up to 1973 when Weihs quantified rapid starting and found maximum acceleration rates of 40–50 $m\,s^{-2}$. The main effort to advance this branch of science was made by Webb with nine papers between 1975 and 1986. The results obtained were reviewed by Harper and Blake (1990). These early studies use over-smoothed data, either from curves fitted to distances obtained at low frame rates, or from double differentiation of distance–time data. As a result, the maximum acceleration values from earlier studies are low. Harper and Blake (1990, 1991) combined high-speed film techniques with direct measurements of acceleration, using subcutaneously implanted accelerometers, of rainbow trout and northern pike, *Esox lucius*. They startled the fish in the experiment by thrusting a wooden pole at the head of the fish.

Weihs (1973a) described the fast start in three successive kinematic stages. A preparatory stage, in which the fish bends its straight body into a curve, is followed by a propulsive stage, and the start ends with a variable stage, where a fish may glide or continue to swim. Webb (1976) distinguished between S-shape and C-shape body curves during fast starts. Webb (1978) describes how the median fins are erected to provide a large area interacting with the water. The paired fins are kept adducted. Harper and Blake (1990) differentiate among three types of startle response; types 1 and 2 are used by trout and pike and type 3 exclusively by the pike.

Fig. 6.2((a) to (c)) represents kinematic data for the three fast-start types of pike. A fish using the first type bends its body into an S-, then a C-shape and turns considerably more than 90° from its original track. The acceleration increases to one significant peak when the head is pointed in the direction of the escape or attack.

In the second type of startle response, the fish bends in quick succession into an S-, a C- and another S-shape. There are two peaks of acceleration, one during the preparatory stage, the other when the C-shape turns into an S-shape again. The fish turns about 90° and the velocity peaks twice.

The sequence of bending in the third type of response is similar to that in type 2. The main difference from types 1 and 2 is the final direction of movement, which is the same as the original direction. Three acceleration peaks occur.

In a subsequent paper (Harper and Blake, 1991), fast starts of pike during feeding were shown to be different from the three types of fright response. A feeding pike usually strikes in the direction of the original orientation of the body. The body bends into an S-shape and a fast wave of curvature starts to run down the body, resulting in a number of half-tail strokes. If the prey is very close, there is only one half-stroke of the tail and the direction of the attack is often at a small angle relative to the longitudinal axis at the start. There is one acceleration peak, with a maximum occurring when the head is orientated in the direction of the prey. The number of half-strokes increases with the distance between the pike and its prey. A distant prey is always attacked straight ahead without changes of direction. A strike at a distance of slightly more than one body length required four half-strokes and showed four peaks of acceleration. The rate of acceleration decreased with strike distance.

Table 6.2 compares the maximum values found by Harper and Blake (1990, 1991) for the startled rainbow trout (L = 0.37 m, mass 0.49 kg) and the escape and feeding reactions of its potential predator the northern pike (L = 0.38 m, mass 0.38 kg). The pike is obviously a specialist with maximum acceleration rates up to more than twice the maximum values for trout. We have to keep in mind, however, that the trout tested here came from a hatchery; wild trout might perform a lot better. The feeding pike's

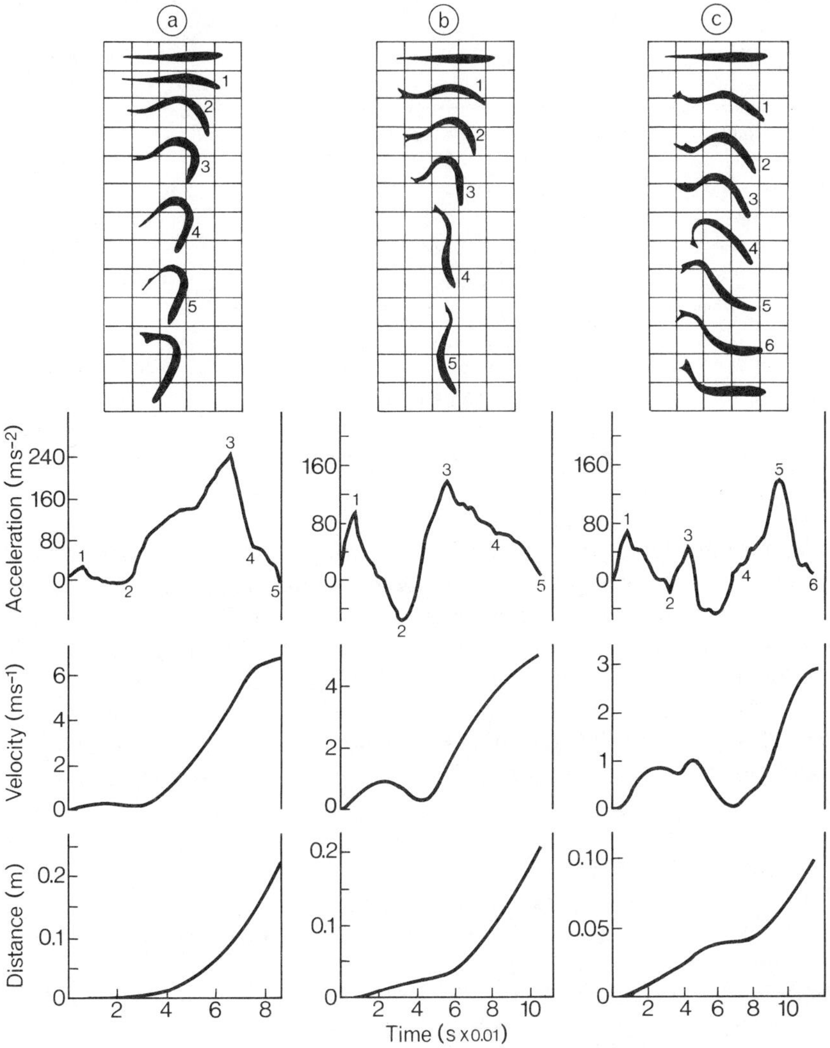

Fig. 6.2 (a) to (c) Kinematic data for three types of fast start of the pike, *Esox lucius*, startled by thrusting a pole at the head to obtain maximum performance. The fish drawings have been displaced sideways and downward over arbitrary distances for greater clarity. The numbers in the drawings correspond with those in the acceleration plots. Redrawn with permission from Harper and Blake (1990).

Table 6.2 A comparison of the startle performances of a predator and its prey. Based on data collected by Harper and Blake (1990, 1991)

Species	*Behaviour*	*Distance* (*m*)	*Maximum velocity* (*m* s^{-1})	(L s^{-1})	*Maximum acceleration* (*m* s^{-2})
Oncorhynchus mykiss	Escape				
	Type 1	0.149	4.19	11.3	95.7
	Type 2	0.223	3.94	10.6	97.8
Esox lucius	Escape				
	Type 1	0.213	7.06	18.8	244.9
	Type 2	0.289	4.70	12.5	141.2
	Type 3	0.227	4.50	12.0	130.5
	Feeding				
	Type 1	0.158	3.35	8.8	159.6
	Type 2	0.221	3.80	10.0	118.6
	Type 4	0.311	4.61	12.1	136.0
	Type 4	0.429	4.75	12.5	72.9

need for directional stability probably prevents the use of the highest acceleration rates possible. The maximum acceleration rates of the feeding pike are higher than those of the escaping trout for strike distances shorter than one body length. The body shape, the positions of the unpaired fins, the predominantly white musculature and one extra fast-start technique are all adaptations to the ambush predator niche of the pike. The rainbow trout has the characteristics of a stayer, but is nevertheless still capable of short bursts of accelerations reaching 10 g (g is the acceleration due to gravity of 9.8 m s^{-2}). The maximum g value measured for pike was almost 25. A space flight rocket launch exposes astronauts to peak values of about 10 g. A record tolerance of 31 g for 5 s has been established for a completely submerged person (Bullard, 1966).

The sequence of events during the startle response of the angelfish is different from that of the three types found for trout and pike, because it starts with a C-shaped bend of the body and not with an S-shaped bend (Domenici and Blake, 1991). Kinematic analysis (Fig. 6.3 (a) and (b)) of angelfish (L = 7.26 cm, mass 8.55 g) shows that the subsequent movements can be classified as a single-bend movement in which the body and tail quickly straighten, starting from the C-bend, or as a double-bend start. The double-bend start shows a full return flip after straightening out of the C-shape. The single-bend start shows one peak of acceleration during the first 16 ms when the body bends into the C-shape. The double-bend start is characterized by two large acceleration maxima, one after about 10 ms and the other after 20 ms. The turning angle of the single bend is about 120°

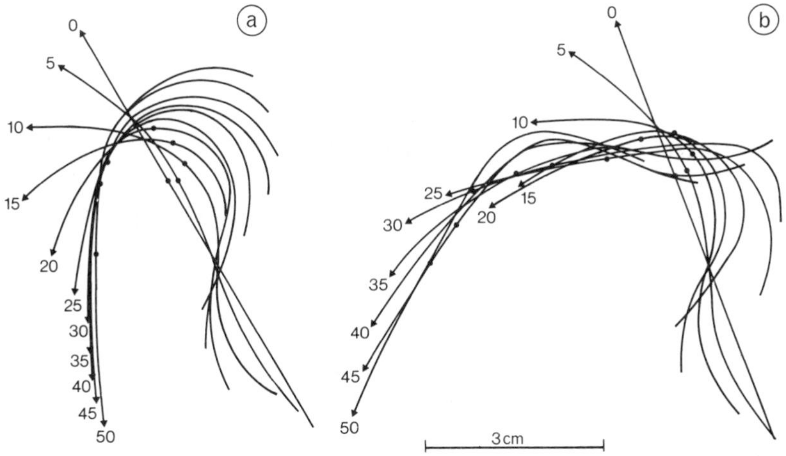

Fig. 6.3 Tracings of a single-bend start (a) and a double-bend start (b) of the angelfish. The midline of the fish in dorsal view and the centre of mass (black dots) are shown. Numbers indicate time (ms). Reproduced with permission from Domenici and Blake (1991).

and that of the double, approximately 73°. The maximum acceleration values of the angelfish and trout are similar. The distance covered and the maximum velocities are in the same order of magnitude if compared in terms of body length.

Rapid turns

Quick, tight turns are characteristic features of fish manoeuvring skills, but few studies have measured how tightly fish can actually turn. The average radius of the turning circle of the angelfish, measured by Domenici and Blake (1991), is only 0.065 *L*. Webb (1983) measured turning radii of 0.18 *L* and 0.11 *L* for rainbow trout and smallmouth bass, *Micropterus dolomieu*, respectively. The minimum turning radius of a fish depends on the body flexibility, the degree of lateral compression, and the lateral area to generate thrust. The radius is independent of speed and acceleration. A small turning radius can be beneficial to both sides in predator–prey interactions, but it is also very useful in a complex environment such as a coral reef or a densely vegetated freshwater system.

Braking

The intentional retardation of forward movement, as Aleyev (1977) defines braking, has received little attention despite its ecological importance.

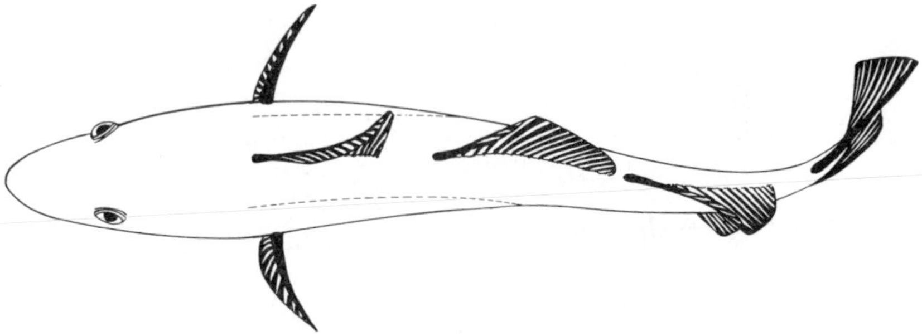

Fig. 6.4 Drawing from a picture of a braking cod (L = 0.42 m; mass 0.67 kg). Reproduced with permission from Videler (1981).

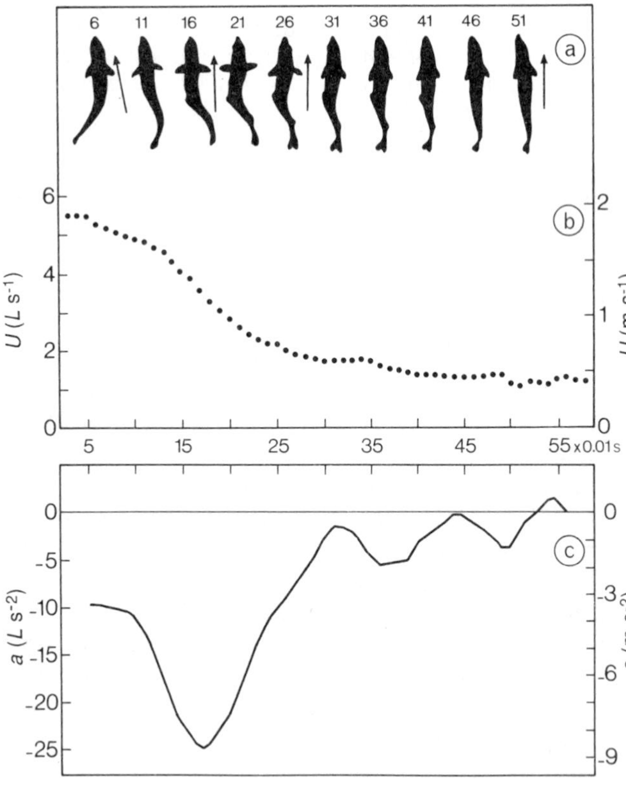

Fig. 6.5 Kinematic analysis of a high-speed film of a braking saithe. (a) Outlines of the dorsal view of the animal at the frame numbers indicated. (b) Velocity profile as a function of time (indicated as frame numbers). The frame rate was 100 Hz. (c) Deceleration as a function of time. Reproduced with permission from Geerlink (1987).

Demersal fish living in a complex environment rely strongly on their manoeuvrability, which includes the ability to brake. Some fish have to brake for a living. The greater forkbeard, *Phycis blennoides*, for example, feeds on bottom-dwelling shrimp. While feeding, this species swims fast and close to the bottom with the extremely elongated pelvic fins extended laterally for the detection of prey. When the forked pelvic fins touch a shrimp, the fish instantly spreads the long dorsal and anal fins and throws its body into an S-shape. Braking is so effective that the shrimp has not yet reached the caudal peduncle before the fish has stopped and turned to catch it (this description is based on a research-film made by Dr G. Thomas).

Fish use the unpaired fins and tail, usually in combination with the pectoral and pelvic fins, for braking (Harris, 1937). In the process, the fin rays of the tail fin are actively bent forwards (Bainbridge, 1963). McCutchen (1977) filmed a fish braking with pectoral fins and a puff of water out of the mouth, suggesting an additional braking effect from this inverse jet. The cod of Fig. 6.4 (Videler, 1981) was forced to brake while swimming at a velocity of 1.26 $m\,s^{-1}$ and decelerated at 2.3 $m\,s^{-2}$. Geerlink (1987) measured deceleration rates during voluntary braking of cod (L = 0.26 m, mass = 0.18 kg), mackerel (L = 0.34 m, mass = 0.34 kg) and saithe (L = 0.35 m, mass = 0.57 kg). The highest deceleration rates were 1.7 $m\,s^{-2}$ for cod, 3.7 $m\,s^{-2}$ for mackerel and a maximum value of 8.7 $m\,s^{-2}$ for saithe. Fig. 6.5 shows the kinematic details of this extreme action of saithe. Geerlink (1987) estimated the contribution of the pectorals to the braking force to be about 30%, the rest being from the curved body and extended median fins.

Burst-and-coast swimming

Burst-and-coast (or kick-and-glide) swimming behaviour is commonly used by several species. It consists of cyclic bursts of swimming movements followed by a coast phase in which the body is kept motionless and straight. The burst phase starts off at an initial velocity (u_i), lower than the average velocity (u_c). During a burst the fish accelerates to a final velocity (u_f), higher than u_c. The cycle is completed when velocity u_i is reached at the end of the deceleration during the coast phase. Fig. 6.6 shows the velocity profile of this swimming style for a 0.26 m cod (Videler and Weihs, 1982).

Energy savings in the order of 50% are predicted if burst-and-coast swimming is used instead of steady swimming at the same average speed, during slow swimming (Weihs, 1974) and for high swimming speeds (Videler, 1981). The model predictions are based on a substantial difference in drag between a rigid body and an actively moving fish. To swim at a certain average speed, a fish could chose a variety of u_i and u_f values. Our model (Videler and Weihs, 1982) predicted the combination of $\mathrm{u_i}$ and $\mathrm{u_f}$ that could

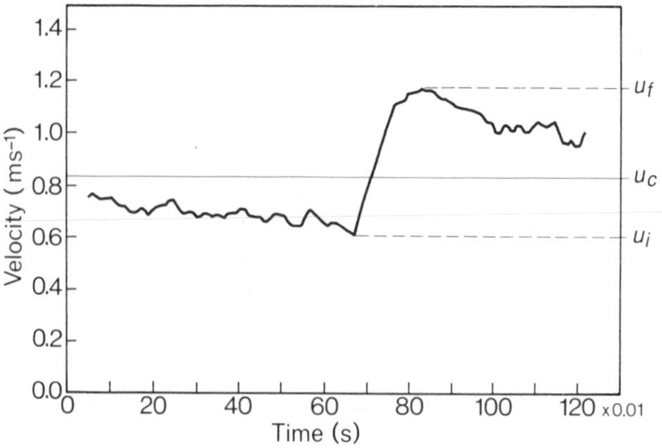

Fig. 6.6 Part of the velocity curve during burst-and-coast swimming at an average velocity (u_c) of about 3.2 $L\,s^{-1}$ of cod; u_f is the final speed, u_i the initial speed during the acceleration phase. Reproduced with permission from Videler and Weihs (1982).

save as much energy as possible for a range of average speeds. If we assumed (based on Videler, 1981) that the drag on a steadily swimming fish was three times the drag on a gliding fish at the same average speed, we could demonstrate that cod and saithe make use of the predicted advantage by choosing the initial and final burst velocities close to the optimal values.

Not all fish species are designed to make optimal use of the energetic advantages of burst-and-coast swimming. Bodies with a fineness ratio (body length over largest diameter) of 5 are optimal (Blake, 1983b). Frequent burst-and-coast swimmers (e.g. Gadidae, Clupeidae) are characterized by fineness ratios of 4.0–6.5.

Tilting

At very low speeds some fish species swim horizontally with the body axis not kept parallel to the horizontal. This 'tilting' behaviour was originally attributed to negatively buoyant fish. He and Wardle (1986) described how Atlantic mackerel (a species without a swim bladder) swims head-up (or tail-down) with extended pectorals at speeds between 0.3 and 0.8 $L\,s^{-1}$. They tried to explain this behaviour by pointing at the need of these fish to generate dynamic lift to balance the weight. The downward-pointing tail blade generates thrust in an obliquely upward direction, contributing to vertical stability. Apart from that effect the area of the lifting surfaces is increased by upward tilting, so lift from the body is added to that of the pectorals. The lift-generating capacity of the extended pectoral fins is strongly reduced at low speeds.

Webb (1993a) showed that nearly neutrally buoyant fish species may also use this behaviour, and pointed at the fact that tilting is a widespread phenomenon among slow-swimming fish. Trout and bluegill swam with the body tilted under experimental conditions, trout, like mackerel, with a positive tilting angle at swimming speeds between 1 and 2 $L\,s^{-1}$, and bluegill with a positive or negative (head-down) tilting angle at velocities between 0 and 2 $L\,s^{-1}$. Webb suggests that tilting with positive or negative angles can be an active mechanism in the stability control of all fish, either negatively or neutrally buoyant. Tilting, plus the increased extension of fins for passive control at low speeds, will increase drag. Propulsive surfaces will have to generate more thrust and hence provide larger forces for stability control without increasing the overall speed.

Stability is a typical problem related to low velocities for animals and vehicles operating in water or air. Tilting is probably the behaviour used by many fish species to overcome that problem.

6.4 PRECISE KINEMATICS OF THE TAIL

So far we have been treating the caudal fin as a curved, flat plate extension of the main body, implying that only one estimate for the angle Θ between the fin and the mean path of motion in the horizontal plane was made for each frame of film. That is not correct, because fish tails do not behave as flat plates and therefore different parts of the fin make different angles Θ with the swimming path. This undoubtedly will have a large effect on our estimates of the forces generated by the tail. Bainbridge (1963) studied the detailed movements of the caudal fin of dace, goldfish and trout at close range during series of tail beat cycles and I did the same for tilapia, (Videler, 1975). In the horizontal or frontal planes, anteroposterior bending determines to a large extent the value of Θ near the end of the fin. This bending is directly related to the individual stiffness of the fin rays. We saw in Chapter 2 how fin ray stiffness can be modulated by contraction of the intrinsic muscles inserting on the left and right side of the fin ray heads. The longitudinal stiffness of the rays near the dorsal and ventral edge of the fin is usually structurally different from those in the centre of the fin.

In the transversal vertical plane through the end of the fin we can obtain a clear caudal view of the dorsoventral curvature of the fin (Fig. 6.7). During steady swimming the dorsal and ventral margins lead the lateral movements and the centre lags behind. But that is not always the case. Tilapia, for example, uses for some manoeuvres its intrinsic muscles of the tail fin to make the centre, or either the dorsal or ventral edge, lead the stroke. The height of the end of the tail is affected by the amount of dorsoventral curvature and by the spreading actions of the fin rays. Thomson (1976)

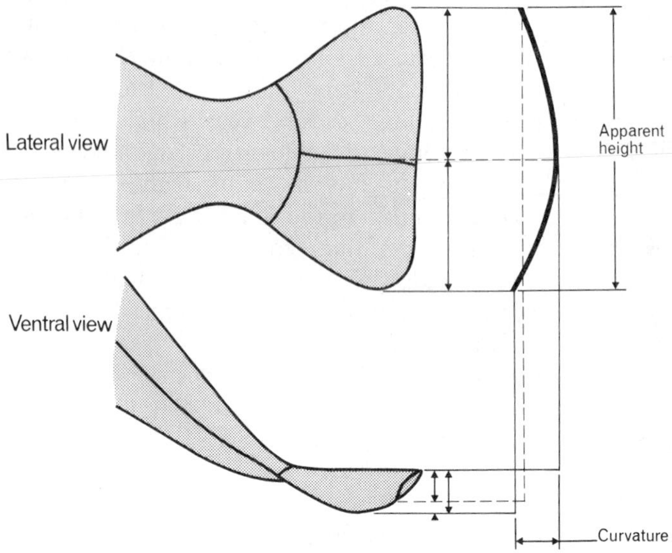

Fig. 6.7 Reconstruction of the caudal view of the tail blade of *Sarotherodon niloticus*, using simultaneous pictures of the lateral and ventral side. The apparent height and a measure of the curvature are defined. Adapted from Videler (1975).

studied motion pictures of the tails of sharks, taken from directly behind free-swimming fish, and found similarly different patterns of dorsoventral and longitudinal bending.

Spreading determines the surface area of the fin in the median plane. Bainbridge (1963) found rhythmic changes related to the tail beat cycles for the dorsoventral curvature and the height of the caudal fin. Fig. 6.8 uses the data for dace to illustrate that the height variation was out of phase with the curvature. The maximum height occurred just after the tail tip crossed the mean path of motion in the middle of the stroke. At these instants the curvature is increasing from the minimum values that it showed a fraction earlier during the same stroke. Maximum curvature values occur just before the tail tip reaches the lateral extremes of the transverse movement while the height is decreasing. The height changes are twice as large as could be explained from changes in height owing solely to dorsoventral curvature. These phenomena can only be explained if the fish is actively spreading and closing the fin cyclically.

The upper lobe of the tail of mackerel leads in the direction of the lateral tail sweep at low speeds, lifting the rear of the body (He and Wardle, 1986). This becomes less obvious at higher speeds.

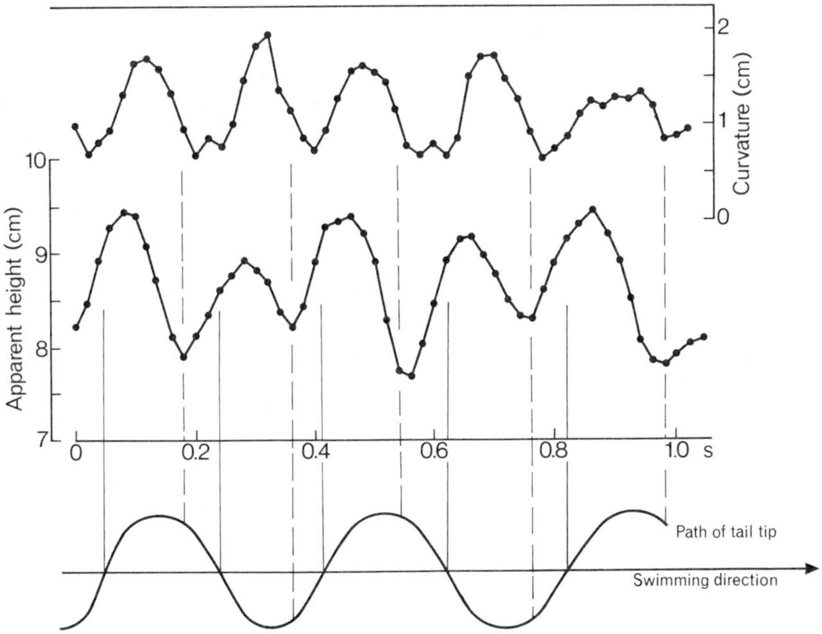

Fig. 6.8 Rhythmic changes during tail beat cycles of the apparent height and curvature of the tail blade of a 30 cm long dace, during steady swimming at 0.48 $m\,s^{-1}$. Curvature and apparent height are defined in Fig. 6.7. Adapted from Bainbridge (1963).

6.5 SWIMMING WITH APPENDAGES

Representatives of at least 12% of the 450 extant fish families (Nelson, 1984), do not use lateral movements of body and tail as their routine propulsive technique. Many more species use paired and unpaired fins for manoeuvring and stability, especially at low speeds. Many paired and median fin swimmers start to use body and tail movements to reach the highest velocities during escape reactions. Despite the relatively common use of appendages in fish swimming, kinematic studies are rare.

Pectoral fin swimming

Pectoral fin movements during straight, forward swimming in a water tunnel by the shiner perch, *Cymatogaster aggregata*, were studied by Webb (1973). This species swims exclusively using the pectoral fins at speeds between 0.5 and 3.4 $L s^{-1}$. The beat cycle consists of three phases. During the abduction phase, the dorsal fin rays lead the movement of the fin away from the body and slightly downward. The ventral side of the fin trails behind. The adduction phase brings the fin back to the body surface, led by horizontal movement of the dorsal fin rays. During the third or intermediate phase of the cycle, the dorsal rays rotate over an angle in dorsal direction to bring the tip of the leading edge back to the initial position against the body. With increased swimming speeds the duration of the intermediate phase decreases from 65% of the cycle time at 1 $L s^{-1}$ to less than 10% at 4 $L s^{-1}$. The ab- and adduction phases take the same proportion of the cycle period, both increasing proportionally with speed. The length of the wave on the fin increases with swimming velocity. At slow

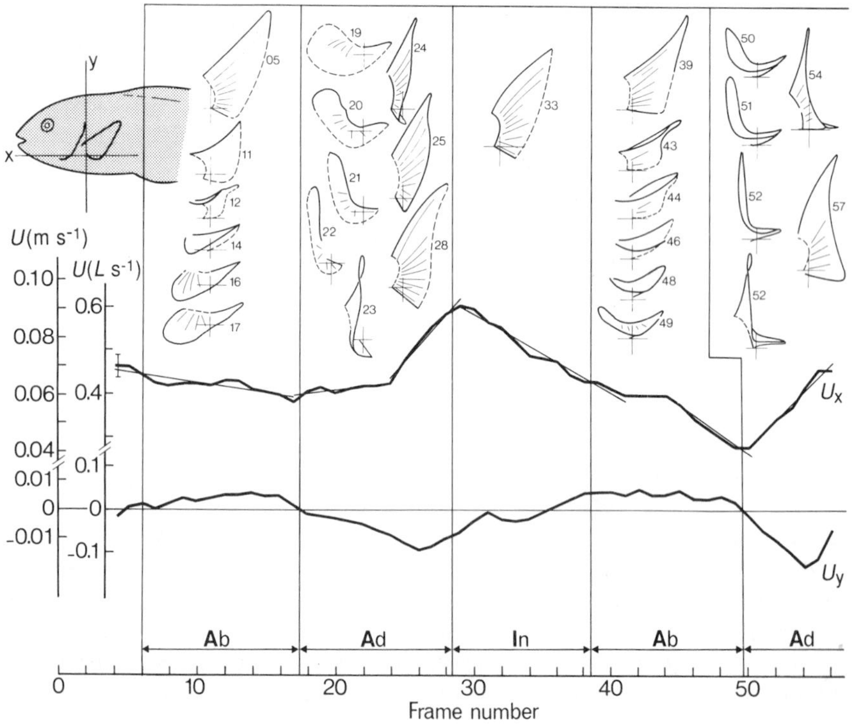

Fig. 6.9 The left pectoral fin beat cycles of *Coris formosa* (Labridae) (L = 15 cm), during voluntary slow forward swimming. The left and right pectoral fins were beating symmetrically. Time is indicated as frame numbers; the frame rate was 100 Hz. Ab, abduction phase; Ad, adduction phase; In, intermediate phase. Frame numbers are indicated near each fin drawing. The straight line-segments in the U_x curve give an indication of phases in the finbeat cycle showing uniform acceleration or deceleration. After Geerlink (1983).

velocities the rear edge trails about π behind in phase with the leading edge. At higher swimming speeds this phase delay reduces to values of about 0.2 π. Both frequency and amplitude increase in a non-linear way over the total speed range of 0.5 and 3.5 Ls^{-1}, where the animals use the pectorals exclusively for propulsion. The product of amplitude and frequency increases linearly with speed. The shiner perch reaches very high stride length values from 0.67 L at 1 Ls^{-1} up to 1.14 L at 2.5 Ls^{-1}, decreasing while swimming faster to 1 L at 4 Ls^{-1}.

The pectoral fin beat cycle of *Coris formosa* (Geerlink, 1983) is similar in that it shows the same three phases. Geerlink measured instantaneous velocities during voluntary unrestrained swimming in an aquarium with static water. These conditions offer slow swimming with large variations between the cyclic events. Fig. 6.9 is an example of the effect of the pectoral fin beat cycle on the instantaneous velocity of the fish (measured at a brightly coloured point on the head), in the swimming direction and also

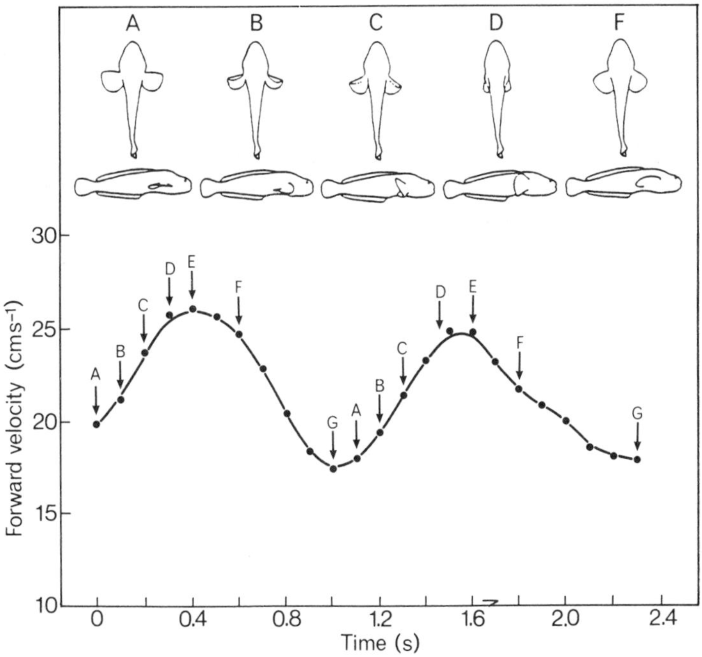

Fig. 6.10 Pectoral fin swimming at 2 °C of *Nothothenia neglecta*. The forward velocity over two pectoral fin beat cycles, and a diagrammatic representation of the fin beat pattern are shown. A–D, adduction phase; D–E, intermediate phase; E–F, initial part of abduction phase when pectorals are spread; E–G, gliding period; F–G, gliding phase with extended pectorals; G–A, final part of abduction phase. Modified after Archer and Johnston (1989).

in the vertical plane. There is a speed increase during the adduction phase coinciding with a downward movement of the head. The head moves upward during abduction. The stride length was 0.16 L in this case, which is much lower than the values found for the shiner perch.

Pectoral fin movements usually make an elegant impression owing to the waves that run down the fin during the stroke. The subtlety of the effects of pectoral fin swimming movements is nicely described for the Antarctic fish, *Notothenia neglecta*, swimming at 2 °C, by Archer and Johnston (1989). They trained these fish to swim between feeding lights along a 4 m long raceway. Adult *Notothenia* used only the pectorals to swim at velocities lower than 0.8 $L s^{-1}$, while juveniles used pectorals alone to 1.4 $L s^{-1}$. Fig 6.10 illustrates two fin beat cycles for an adult. The speed increases during the adduction phase. In this phase, successive fin rays move both laterally backward and upward, producing a sinusoidal wave over the fin. It is followed by an intermediate phase where the leading rays move dorsally against the side of the body. Abduction occurs in three stages. Initially the fin is spread to form a pair of wings on which the negatively buoyant fish glides and decelerates. Just prior to adduction the leading edge completes abduction by moving further forward and ventrally. At an average speed of 0.8 $L s^{-1}$, the mean stride length is 0.77 L and the fin beat frequency about 1 Hz.

Skates and rays use their broad pectoral fins to fly underwater. The slow flight-like appearance makes swimming of e.g. large devil rays (Mobulidae) or eagle rays (Myliobatidae), extremely spectacular to watch. Daniel (1988) analysed the kinematics and dynamics of the forward flapping aquatic flight of the clearnose skate, *Raja eglanteria*. This species swims with waves on the pectoral fins running backward at approximately twice the forward velocity. The amplitude is fairly constant in the swimming direction, but increases drastically in lateral direction towards the fin tips.

Median fin propulsion

A series of studies on locomotion with undulations of dorsal and/or anal fins has been published by Blake between 1976 and 1983. The prime intention of these papers was to estimate propulsive force and power; kinematic descriptions were of secondary importance. The following paragraphs summarize the scant data.

Sea-horses (Syngnathidae) hardly swim at all. Their swimming capacity is reduced to turning and minute displacements forward, backward and up and down, using their dorsal and pectoral fins. Breder and Edgerton (1942) measured a dorsal fin beat frequency of 35 Hz and noted that the sea-horse, *Hippocampus hudsonius* could change the wavelength and amplitude. There were between 1.3 and 2.5 complete waves on the length of the dorsal fin. Blake (1976) found similar dorsal fin frequencies for an 11 cm long specimen of the same species, swimming at 5.5 $\mathrm{cm\,s^{-1}}$.

Boxfishes (Ostraciontidae) use waves on the dorsal and anal fin as well as on the pectoral fins for routine swimming. The tail usually acts as a rudder. These fish swim forward, backward, turn, bank and even swim upside down. Blake (1977) gives highest values for the frequencies of the waves on the dorsal, anal and pectoral fins of 3.8, 4.5 and 2.8 Hz respectively, for a 12 cm *Tetrasomus gibbosus* swimming at 11.2 $\mathrm{cm\,s^{-1}}$. At this high speed, tail beats aid propulsion by left-to-right movements over an arc of 70°. The tail blade is spread and accelerates from one extreme left or right position to the median position and decelerates from there to the extreme position on the other side.

Triggerfishes (Balistidae) swim predominantly with undulations on their dorsal and anal fins, although pectoral fins are frequently involved in slow swimming and manoeuvring and the tail aids with fast swimming. The undulating waves on the median fins pass from front to rear during forward swimming, but the direction can be reversed for braking or backward swimming. Blake (1978) shows that the frequency of the waves on the dorsal and anal fins of the Picasso triggerfish, *Rhinecanthus aculeatus*, increases linearly with swimming speeds above approximately 0.5 $L\,\mathrm{s^{-1}}$. At speeds lower than that and during hovering, the frequency is higher than the minimum value at 0.5 $L\,\mathrm{s^{-1}}$. The stride length is somewhere between 0.25 and 0.3 L. Blake's illustrations of the waves on the dorsal and anal fin of the Picasso triggerfish, swimming at 5.1 $\mathrm{cm\,s^{-1}}$, reveal that there is 1.5 wavelength on each fin at each of the four moments in time shown. These diagrams allow calculation of the wave speed, which turns out to be 3.7 $\mathrm{cm\,s^{-1}}$ for both fins. Blake did not mention wave speeds and was obviously not aware of these low values. It shows that he filmed a fish that was trying to delay its progress by sending waves down its fins in the right direction but at a slower speed than the swimming speed at that instant. A graph of swimming speed against time as in Figs 6.9 and 6.10 would have shown this braking effect. This discovery provides insight into the sophistication of this propulsive system which is probably an adaptation directed at precise and efficient manoeuvring. The high efficiency of a swimming apparatus with a large rigid body, propelled by waving fins (see also Section 4.4), arises because side forces are minimized (thereby reducing yaw), and because most of the thrust can be directed backward (Lighthill and Blake, 1990).

The American knifefishes and electric eels (superfamily Gymnotidae) and the African and Asian featherbacks (family Notopteridae) swim by undulations of the extremely long anal fin. Blake (1983a) measured the length of the anal fin base of six species of Gymnotidae. The average length was 73.7% of the body length, varying between 65.4% and 82.5%. The anal fin lengths of the three species of Notopteridae were all just over 81% of the body length.

Analogously, the African electric eel, *Gymnarchus niloticus*, uses undulations of the long dorsal fin which extends over almost 70% of its body length and contains between 183 and 230 rays (Nelson, 1976). The number of complete wavelengths on these anal and dorsal fins varies between 2.5 and 1.5. The waves move equally well forward and backward.

The variation in swimming kinematics among fishes is larger than the number of species. The total number of species that has been investigated properly is extremely limited. Reviewers after Breder (1926) usually searched for general rules using a very limited data set. I think that in this case, the diversity is much more interesting than generalizations. From a biological point of view, insight into the diversity of swimming styles helps to understand adaptive radiation among related species and shows analogous adaptations to specific niches. From a biomechanical standpoint, explaining the purpose of the kinematic diversity is a greater challenge than unifying it into a few approximate models.

6.6 SUMMARY AND CONCLUSIONS

A compilation of kinematic data on swimming at uniform speeds without the use of pectoral fins reveals some trends explaining large proportions of the total variation. The average tail beat amplitude is close to 10% of the body length. The average stride length is about 0.7 *L*. The ratio of forward speed over body wave speed backwards tends to be closer to 1 (less slip) when the strides are longer.

Specific differences between kinematic data can be fairly large. This is probably partly due to variation in the methods used, but also caused by real differences in swimming skills usually related to a specific ecological niche. Pike use smaller tail beat amplitudes and higher frequencies than trout, marking the difference between a fast starter and a stayer. A number of specific differences are discussed.

The maximum stride length is approximately speed and length independent within the normal range of adult velocities, but it varies between species. Larval stages have shorter stride lengths.

Unsteady swimming is more common than swimming at uniform speeds but more difficult to study kinematically. The sequences of kinematic events during different types of fast starts are described. Evidence reveals that rapid-starting fish can achieve extremely high acceleration values. The maximum value published for pike is nearly 25 *g*.

The kinematics of rapid turning, braking and burst-and-coast swimming are described. The minimum turning radius depends on the flexibility of the body, the degree of lateral compression of the body and on the lateral area of the propelling surface. Braking is an important skill, both for the acquisition of food and for collision avoidance. The highest deceleration value measured was 8.7 $m\,s^{-2}$ for saithe. Intermittent swimming behaviour saves energy if the difference in drag between active swimming and gliding with a rigid body is large enough. There are no exact data, but estimates point in the direction of three times more drag during swimming than during gliding.

Tail fins of fish are not simple flat plates at the end of the body. Fish are in control over the bending properties of each fin ray and hence can influence the angle between the end of each part of the tail blade and the mean path of motion. Fish cyclically change the surface area of the tail blade during successive swimming strokes.

A compilation of kinematic data on fish swimming with the pectorals and of various groups using median fins for propulsion, concludes this chapter.

Chapter seven

Swimming dynamics: work from muscles

7.1 INTRODUCTION

The propulsive movements of fish are powered by muscles, either by the intrinsic muscles moving fins or by the lateral fibres arranged in myotomes on both sides of the body axis. The relation between form and function of these last muscles deviates from that of other skeletal muscle systems in vertebrates, because they do not run between an origin on the skeleton and an insertion across a joint somewhere else on the skeleton. Red and white fibres of the myotomes are used by fish to generate propulsive undulations of the body and we would like to know exactly how. The biomechanics of the antagonistic abductor and adductor muscle systems of fins appear to be less complicated. We shall therefore focus our attention to the lateral muscles in the next chapters. The physiological properties of fish muscles are the subject of this chapter, and how they propel the fish is discussed in Chapter 8.

We will start at the sarcomere level, continue with an overview of studies on the performance of bundles and blocks of fibres *in vitro* and finish this chapter by discussing electromyography, a technique that can be used to study muscle function *in vivo*.

At the sarcomere level, fish muscles do not differ from frog muscles, where the amount of overlap between thick and thin filaments determines the sarcomere length and the magnitude of the force in an isometric contraction. Sarcomere length excursions of red and white carp muscle fibres during steady swimming and escape responses have been determined and are shown to be close to optimum values.

Force–velocity curves of bundles of muscle fibres provide insight into the velocity range where the highest power is generated. Measurements of these curves in dogfish and carp exemplify the performance of lateral muscle fibres during isotonic contractions where muscles generate force while shortening at a constant velocity.

During swimming, lateral muscle fibres will perform cyclic length changes and are activated during part of each cycle. The forces generated are strongly influenced by the magnitude of the length change and by the instant, duration and strength of the stimulus during the lengthening–shortening cycles. Force times change in length represents the work done by the muscle. For each cycle, a work loop can be constructed which reveals the total amount of positive or negative work done. Optimum conditions for lateral muscles in fish have been studied with this technique.

Blocks of muscles were stimulated *in vitro* to measure the minimum or twitch contraction time. The method has been used to predict maximum swimming velocities for fish of various lengths and swimming at different temperatures. The temperature of fish muscle affects the power that can be generated. The magnitude of the effect of temperature on the power output is not always the same: it also depends on the temperature to which the fish is acclimatized.

Electromyography offers the possibility of studying muscle activity in a swimming fish. The technique has its restrictions but can be used to discover complex patterns of muscle recruitment during locomotion.

7.2 GENERAL MUSCLE PHYSIOLOGY AND SPECIAL FISH MUSCLES

Chapter 2 showed how, at the level of the sarcomeres, muscles operate by making firm connections through the cross bridges between the thin and thick filaments.

We are used to thinking that cross-bridge formation causes the thin and thick filaments to slide along each other, shortening the sarcomere of the contracting muscle. However, muscles can be active without shortening. Cross-bridge formation also generates force in a muscle that is unable to shorten owing to a heavy load that prevents shortening, or even in a muscle that is being stretched forcefully by an outside load. Therefore I will adopt the term 'active' for muscles producing force. Muscles shortening against a load perform a concentric contraction. This type of contraction is called isotonic if the load, and hence the force produced, is constant. When there is no change in length, the muscle is isometrically active and muscles resisting stretch are involved in excentric activity.

An isometrically active muscle produces the highest force when the length of the sarcomeres is approximately twice the length of the thin filaments (Gordon *et al.* 1966). In practice, this length coincides approximately with the resting length in most muscles (Goldspink, 1977).

Maximum isometric forces in swimming carp

The average length of the thin filaments of carp red and white lateral muscles is about 1 μm, and the thick filaments are 1.5 μm long (page 35) The cross-bridge-free zone in the middle of the thick filaments is about 0.13 μm long. These dimensions are the same in frog muscle and therefore Sosnicki *et al.* (1991) suggest that we can use the relation between sarcomere length and force described by Gordon *et al.* (1966) for frog muscles, for the red and white muscles of carp. Fig. 7.1 shows the result of this suggestion: a graph of the percentage of maximum isometric force for the different sarcomere lengths of carp with the diagrams of sarcomeres at four crucial lengths. The longest sarcomere is about 3.6 μm, including the thickness of one Z-disc of about 0.1 μm. There is no overlap between thin and thick

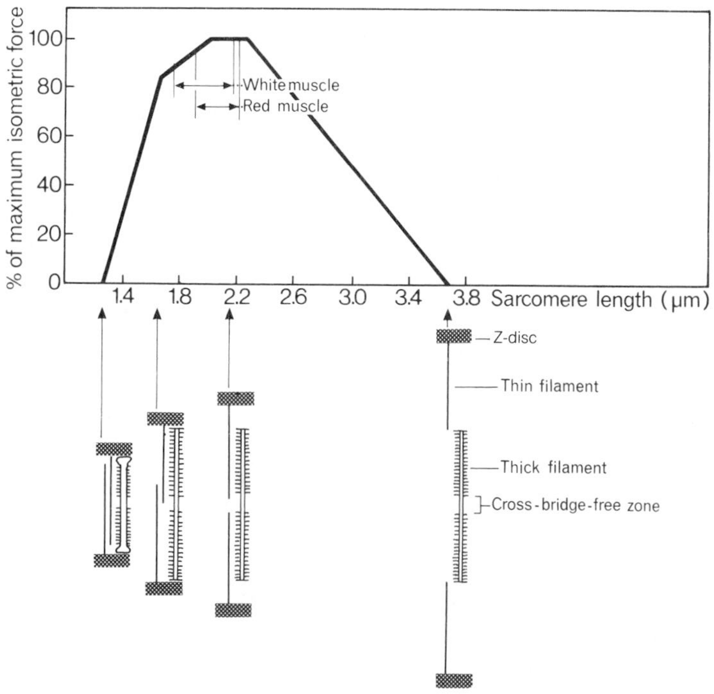

Fig. 7.1 The curve describing the isometric force at a range of sarcomere lengths for frog muscle is used to relate sarcomere extension and relative force in carp red and white muscles. This is a valid approximation because carp myofilament lengths are the same as those in the sarcomeres of frogs. The overlap between thick and thin filaments is illustrated at four crucial sarcomere lengths. The excursion ranges during swimming of carp red and white muscles are indicated. Based on Gordon *et al.* (1966) and Rome and Sosnicki (1991).

filaments, and the isometric force from a sarcomere in this position will be zero. There is also no force generated in the other extreme situation, where the thick filaments abut and buckle against the Z-discs. The distorted sarcomere is 1.3 μm long and the thin filaments overlap completely. The scope for excursion between these two extreme positions is 2.3 μm. Maximum force occurs when all cross bridges are active at sarcomere lengths between 2 and 2.2 μm. Increasing overlap of thin filaments decreases the force up to where the sarcomere approaches the length of the thick filament.

Rome and Sosnicki (1991) found that carp uses its superficial layer of red muscle during steady swimming with sarcomere lengths varying between 1.9 and 2.2 μm, where the force is greater than 96% of the maximum values. At high swimming speeds the deep white muscles take over from the red. The speed at which this occurs depends on the temperature and length of the fish. For a carp of about 18 cm it is at approximately 1.3 $L\,s^{-1}$ at 10 °C and at twice that speed at 20 °C. The excursions of the white fibres during escape responses or high speed swimming will be smaller on average than the changes in length of the red fibres because the white ones are situated more interiorly inside the fish, closer to the vertebral column. However, we saw (page 26) that the white fibres of teleosts are not orientated parallel to the body axis, but are running obliquely in a complex helical way. Alexander's (1969) analysis reveals that this helical arrangement allows the white fibres to shorten over a distance which is approximately equal for all the fibres no matter what their position relative to the vertebral column. Because of this arrangement all the white fibres will only shorten about one-quarter as much as the red fibres for a given change in curvature of the body (Rome *et al.*, 1988).

Sarcomere lengths of white fibres are estimated to vary between 2.2 and 1.75 μm during vigorous left–right escape movements of carp. The isometric tension at these sarcomere lengths would be within 85% of the maximum as indicated in Fig. 7.1. The same changes in curvature would require the red muscle to shorten to about 1.5 μm, where it would only generate 50% of its maximum isometric force.

Change in sarcomere length under locomotory conditions compared with the possibility to generate isometric force show the importance of the overlap between thin and thick filaments in the sarcomere. However, it does not provide insight into forces generated in a dynamic situation where muscles change length at a certain velocity.

Force–velocity and power curves of shortening muscles

An isometrically active muscle generates force but no work, because work is force times distance. A muscle that actively shortens over a given distance at speed V is involved in a concentric contraction.

Concentric contractions allow investigation of the relations between force and shortening velocity. Classical force–velocity experiments have shown how the force generated by a muscle depends on the speed of contraction. A muscle preparation in a physiological experiment will generate a maximum force (P_{max}) if it is activated isometrically. In a completely unrestrained situation the muscle will shorten at its maximum velocity (V_{max}) without exerting any force. In an experiment where one end of the muscle is fixed and the other end attached to a small force and velocity transducer, it becomes possible to measure different velocities for different loads and to construct a force–velocity curve. This sounds easy enough but the actual experimental procedure is rather complicated, especially as far as muscles from fish myotomes are concerned. Many experiments have been conducted on chemically or mechanically skinned muscle fibres of fish. This method was chosen because it produced good repeatable data in cases where attempts to use intact fibres did not give proper results. However, later experiments with intact fibres showed that forces and power measured from skinned fibres could be lower in some species.

Curtin and Woledge (1988) measured force–velocity relationships of bundles of intact white fibres of dogfish and compared the results with experiments on skinned fibres of the same species. I will use their results as an example of the force–velocity approach to muscle mechanics.

Bundles of 1–10 muscle fibres with parts of the myosepts at each end were dissected from the tail region of a freshly killed dogfish. One tendinous end was attached to a force transducer and the opposite end to a motor-driven movable transducer, indicating its displacement in the longitudinal direction of the muscle bundle as a function of time. This preparation was mounted in a bath of saline (Ringer's) solution maintained at 12 °C. Platinum electrodes on either side of the fibre bundle were used to deliver electrical stimuli. Fig. 7.2(a) gives sample records of a test series. Curve (1) shows an isometric tetanus which gives a record of the highest force, P_{max}. A series of measurements, repeated at regular intervals of 3–5 min, were made to establish the force–velocity curve. In each case the bundle was made to contract isometrically to the tetanic force value indicated by the arrow (t) in Fig. 7.2(a). The fibre length was then shortened at constant velocities, cases (2), (3) and (4) in Fig. 7.2((a) and (b)). The shortening velocities are approximately 36, 22 and 11 $mm\,s^{-1}$ for (2), (3) and (4) respectively. During shortening the force dropped at first to reach a plateau value until the change in length stopped. The force value at the plateau was considered to be the force belonging to the particular velocity. In cases (3) and (4), the plateau had time to establish more clearly than in case (2), where it probably only just reached the appropriate value before it started to increase to the isometric tetanic value again.

The highest velocity of the unrestrained muscle was measured in a slightly

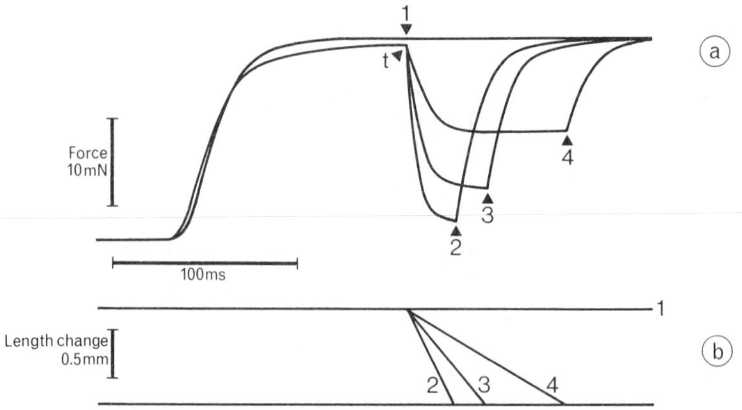

Fig. 7.2 (a) Records of an isometric tetanus (1) and tetani, reached at (t), followed by shortening at constant velocities (2), (3) and (4) of dogfish white muscle fibre bundles at 12 °C. (b) Changes in length as a function of time of shortening of the muscle bundles during tests 1–4 of graph (a). Reproduced with permission from Curtin and Woledge (1988).

different way by releasing the muscle from a tetanus over a series of increasing distances without dictating the shortening velocity. For each distance the force will drop to zero during a short period and then start to develop again. This period increases with increasing release distance. The unloaded shortening velocity is given by the ratio of release distance over the period it takes to start renewed development of tension. Release distances plotted against zero force periods appeared on a straight line for distances between 6.1% and 12.9% of the fibre length. The slope of that line is the maximum unrestrained velocity, V_{max}, over that length range. (This is the 'slack step' method from Edman, 1979.) Results of one experiment for a bundle consisting of eight fibres are illustrated in Fig. 7.3(a), where both force and velocity are expressed as fractions of the maximum values P_{max} and V_{max} respectively. Two hyperbolic functions were fitted. The function represented by the dotted line is forced through the isometric value, (P_{max}, $V = 0$), and through the point V_{max}, where the force $P = 0$. The points for high force and low velocity are above that curve. The solid line, where the point P_{max}, $V = 0$ is omitted, gives a better fit; the y-intercept is much higher than 1. The mechanical power output (the product of force and velocity relative to the maximum value $P_{max}V_{max}$) of the same data points is represented in Fig. 7.3(b). It shows that the maximum power output is produced at shortening velocities between about one-quarter and one-third of the maximum velocity of shortening. The same picture emerges if we look at the combined results for 12 bundles of intact white muscle fibres of the dogfish. Although the scatter is fairly large, the relative power values are generally higher than the

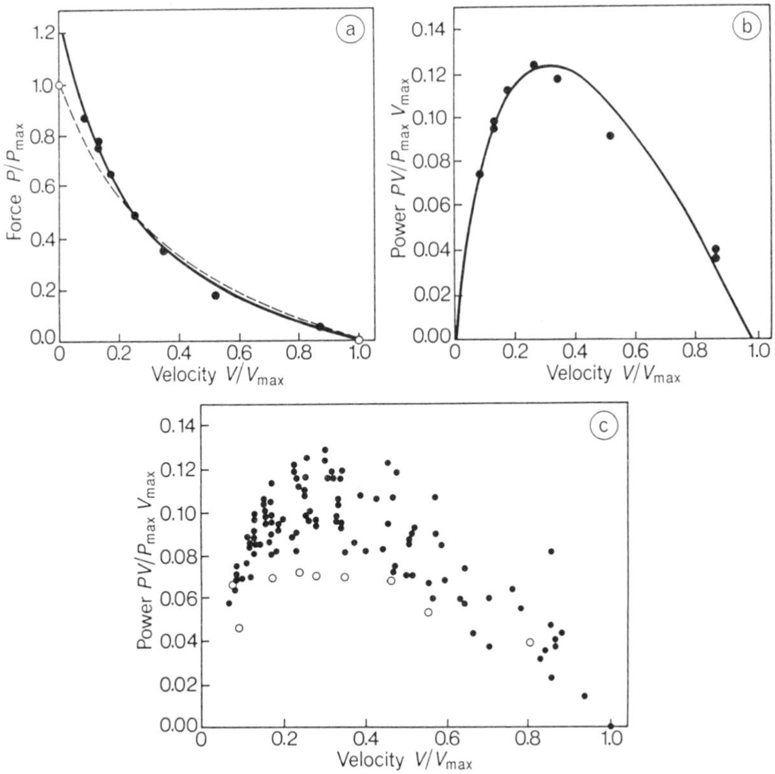

Fig. 7.3 (a) The force–velocity curve of a bundle of dogfish white muscle fibres. The lines are hyperbolae fitted to the data points, either including the value $P/P_{max} = 1$, $V/V_{max} = 0$ (broken curve), or excluding this point (solid curve). (b) The corresponding relationship between the power (force times velocity) relative to the maximum power and the relative velocity of shortening. The curve was calculated from the solid curve in graph (a). (c) Compilation of measurements of power as a function of speed for 12 bundles of intact dogfish white fibres (filled symbols). Results for skinned fibres (open symbols) from Bone *et al.* (1986) are included for comparison. Reproduced with permission from Curtin and Woledge (1988).

data points for skinned white fibres of the same species measured at the same temperature (Bone *et al.*, 1986). The mean value of the maximum power output of intact fibres was 91 W kg^{-1} wet fibre mass, which was about 1.65 times that of skinned fibres. The maximum velocity measured with skinned fibres does not seem to be substantially different from that obtained with intact fibres. The differences in power output are caused by the ability of the intact fibres to produce higher forces at the intermediate velocities. In skinned fibres the chemical medium bathing the filaments is not exactly the same as in live fibres. Although usually great care is taken to mimic the real medium as closely as possible, small deviations may account for the difference.

Rome *et al.* (1988) made force–velocity curves for red muscle bundles in carp over roughly the same length excursions as shown during swimming. The V_{max} at 15 °C was about 4.65 muscle lengths per second. At slow swimming speeds, red muscle in carp shortens at velocities between 0.73 and 1.67 muscle lengths per second. This velocity range is between 0.2 and 0.4 times V_{max}, a range where the muscle generates maximum power. Electromyographic evidence shows that the swimming speed reached with contraction speeds of 1.67 red muscle lengths per second closely approaches the higher speed range where white fibres are recruited to power the movements. A force–velocity curve could not be made for intact white fibres, but skinned white fibres allowed measurement of V_{max} of 12.88 muscle lengths per second. During a startle response, in this case provoked with a sound pulse, the fish quickly bends the body in a C-shape. The shortening velocity of muscle fibres near the surface of the concave side of the fish would have to be 19.6 muscle lengths per second during that movement. This speed exceeds the V_{max} of the white fibres, which are the ones that are active during a startle response, but white fibres are situated closer to the vertebral column and not parallel to the body axis. Owing to this special arrangement they need to shorten at only 4.85 muscle lengths per second to generate the startle response. This shortening speed is about 0.38 times their V_{max}. At that speed they are expected to generate almost maximum mechanical power. Red muscle fibres in the same position as the white would not be able to power the startle response because 4.85 muscle lengths per second exceeds their V_{max}.

Force–velocity curves taught us that muscles generate maximum power during concentric contractions at speeds between approximately one-quarter and one-third of their maximum contraction speed. The behaviour of the muscle fibres in the physiological experiments required to show the relationship between force and velocity is, however, highly artificial. Being forced to shorten at a constant speed starting from a tetanus is not a realistic situation under locomotory conditions inside the animal. Locomotion, and especially fish swimming, is usually a dynamic and rhythmic process where counteracting groups of antagonistic muscles shorten and lengthen in a cyclic way.

7.3 MUSCLE FUNCTION DURING CYCLIC STRAIN CHANGES

Insect flight muscles have little in common with fish myotomal muscles, but both muscle systems produce work during cyclic, near sinusoidal, changes in length under normal operating conditions. Machin and Pringle (1960) emphasized the importance of this fact for the power output of insect muscles and started experiments to quantify this concept. Josephson (1985) extended their method and made it applicable to muscles used during cyclic length changes in general. I will use data from the experiments of Altringham and Johnston (1990a, b) to illustrate the kind of knowledge gained by this approach.

A set of hypothetical curves is drawn in Fig. 7.4 to explain the basic principles. Fig. 7.4(a) shows one cycle out of a series of sinusoidal changes in length of a muscle as a function of time. The horizontal axis gives the cycle duration (the inverse of the cycle frequency), subdivided into 360°. The muscle in this case changes its length symmetrically around its resting length (l_0) from l_0 plus 3% of l_0 to l_0 minus 3% of l_0. The muscle is shortening between 90° and 270° and lengthening during the other half of the cycle. Note that the changes in length are treated separately from the cycles of activation and force generation. These are, as it were, superimposed on the length-change cycles. Changes in length are not necessarily caused by the activity of the muscle itself.

In Fig. 7.4(b), three hypothetical examples are given of force generation (force in N is drawn on an arbitrary scale). In the first example the muscle was stimulated at the onset of shortening, at about 90°. The force develops during shortening to reach a maximum value when the muscle passes through its resting length. The force decreases during the rest of the shortening period. Some force is still present during most of the lengthening phase of the cycle, to drop to a minimum value when the muscle reaches its maximum length. In the second example, stimulation took place at the onset of lengthening at 270°. The force rapidly builds up to an extreme value half-way down the lengthening phase. There is subsequently a fast decrease and little force is generated during the shortening period. The stimulus in the third case arrives at 45° in the middle of the second half of the lengthening period. The force increases rapidly and continues to do so during the first quarter of the shortening period. It drops to a minimum value towards the end of shortening to increase slightly during the first three-quarters of the time used for lengthening.

For each of the cases of Fig. 7.4(b), a work loop is drawn in Fig. 7.4(c). The horizontal axis represents the length (m) relative to l_0, the vertical axis the force (N) on the same arbitrary scale as in Fig. 7.4(b). The thick line from rightmost to leftmost represents the force during the shortening period; the thin line from left to right represents the force during lengthening. The area under the thin line is the amount of work, expressed in N m, that was needed to lengthen the muscle. The area under the thick curve is the work done during shortening. The area within the loop formed by the thick and thin lines represents the net amount of work done during the cycle. In the first case of Fig. 7.4(c), the forces during shortening are higher than during

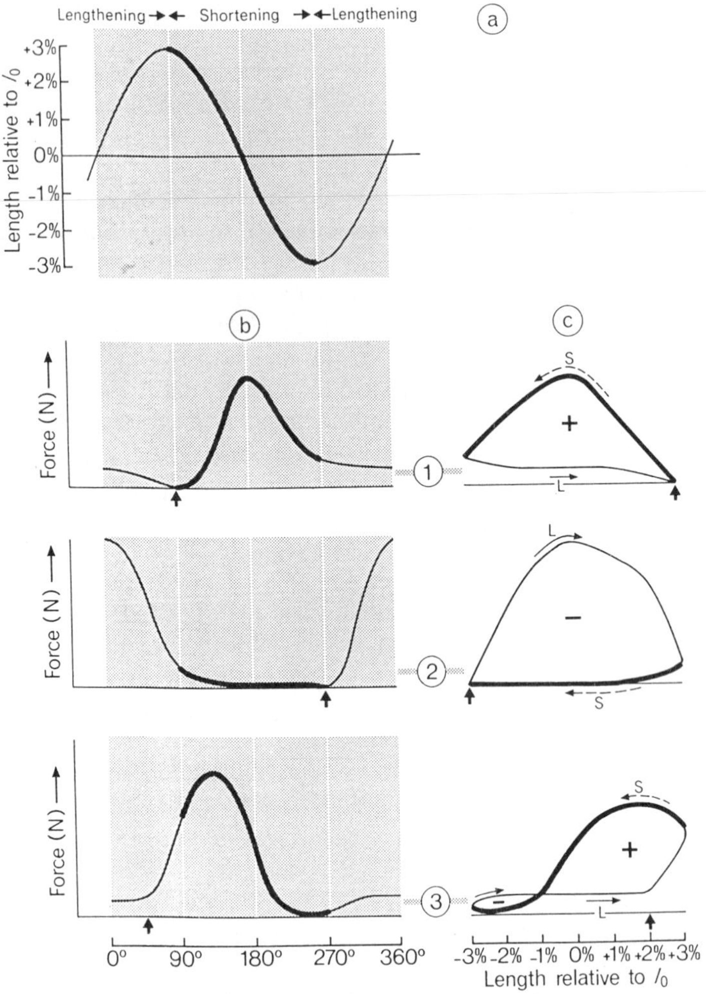

Fig. 7.4 Forces and work loops generated by a muscle stimulated at different phases during sinusoidal changes in length. (a) Relative cyclic length changes around the resting length l_0. One cycle is subdivided into 360°. The shortening phase (indicated by the thick lines in Figs (a), (b) and (c)), starts at 90° and ends at 270°. (b) Three hypothetical cases of force development caused by a stimulus (indicated by arrows) at 90°, 270° and 45° in cases 1, 2 and 3 respectively. The force (N) scale is arbitrary. (c) Graphs of force (arbitrary scale, as in (b)) as a function of muscle length for the cases 1, 2 and 3 of (b). Shortening part of the loop is from right to left (indicated by broken arrows). The lengthening part is from left to right (solid arrows). Work loop 1 is positive because the total positive work in N m during shortening (S) exceeds the negative work during the lengthening (L) period. Work loop 2 is negative and work loop 3 has large positive and a smaller negative part. See text for further explanation.

lengthening and the amount of work done by the muscle is positive. The opposite occurs in the second example, where the largest forces are exerted during the lengthening period. This results in a considerable amount of negative work during that cycle. A hybrid situation is made up as the third case. Forces are large during the beginning of the shortening period but drop to lower values during the beginning of lengthening. The result is a large positive and a small negative work loop. The difference between the two areas gives the net amount of work done during this cycle, which turns out to be positive in this example. Power output of muscles can easily be calculated because it is the net work per cycle multiplied by the cyclic frequency in Hz ($\mathrm{N\,m\,s^{-1} = J\,s^{-1} = W}$).

Work loops of muscle fibres in myotomes

The net work per cycle that a muscle can achieve is clearly not only dependent on the maximum force that can be generated, but also highly influenced by the timing of the active phase of the muscle in relation to the cycle of movement. The relevance to fish swimming movements is obvious. Red and white lateral muscles of fish operate at different frequencies. Each fibre type is therefore expected to have different frequencies for optimal power output. Optimal power output at each frequency will require stimulation at a specific time during each cycle. To find the optimal values, Altringham and Johnston (1990a) tested red and white intact fibre bundles of the bullrout under conditions simulating red and white muscle activity in a fish swimming at different speeds. White fibre preparations consisted of 1–10 fibres, 7.5–14 mm in length; red fibre bundles contained 20–50 fibres, 3–4 mm long. The experiments were carried out at 3 °C with dissected muscle fibres immersed in Ringer's solution. For each experiment one end of the preparation was attached to a servo-motor, the other to an isometric force transducer. A supramaximal stimulus was given during 2 ms via platinum wire electrodes directly into the muscles. The servo-motor induced sinusoidal length changes symmetrical about *in situ* length. The choice of a sine wave was based on the kinematic study of swimming saithe and mackerel (Videler and Hess, 1984) where Fourier analysis had shown that movements of the body during steady swimming are best described as a simple harmonic motion (page 111). Altringham and Johnston stimulated the preparations at selected phases in each cycle and plotted work loops in each case. A test of the effect of the amplitude of the sine wave on power output showed that for both red and white fibres, maxima were reached for amplitudes varying between $l_0 \pm 5\%$ of l_0. Power output turned out to be maximal if the stimulus was given at the end of the lengthening period (at 30° in Fig. 7.4(a)), giving the muscles a small stretch prior to shortening. The higher force during shortening, gained by this timing procedure, outweighed the mechanical costs of stretching active muscle. At high cycle frequencies a single stimulus was sufficient, but at lower frequencies more stimuli were needed to maximize power output.

The results of optimum power output of red and white fibre bundles as a function of cycle frequency are plotted in Fig. 7.5. The insets show the optimum work loops for each point on the white fibre curve; the cycle frequency, number of stimuli and net work are indicated in each case. All the work loops run counterclockwise and are hence positive. The loop at the lowest cycle frequency has the largest area and represents 21.9 μJ of work. The red muscles are slow and develop maximum power output of about 6.5 W kg^{-1} at frequencies between 1 and 3 Hz. The highest optimum power output of the white muscles was 31.5 W kg^{-1} at 5 Hz. The experiments were conducted at 3 °C.

Altringham and Johnston (1990b) did similar experiments at 4 °C with white fibres of cod varying in body length between 13 and 67 cm. The work done per cycle was optimal at length changes of $\pm$ 5% of l_0 for stimuli given at 10° (Fig. 7.4(a)). (Anderson and Johnston, 1992, reported that optimal length changes are size dependent. They found that larger strains, of about 7–11%, were required for maximal power output in cod smaller than 18 cm,

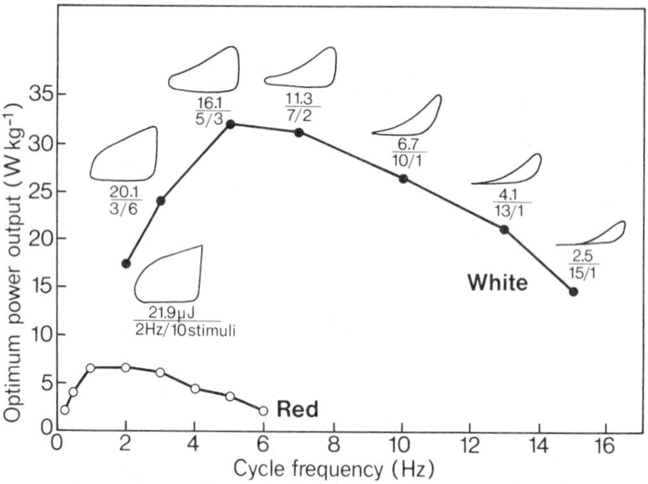

Fig. 7.5 Optimum power output tested at 3 °C plotted against the cycle frequency for white (filled symbols) and red (open symbols) fibres of the myotomes of the bullrout, *Myoxocephalus scorpius*. The optimum work loops are shown for each point of the white fibre curve, with indications of the cycle frequency and the number of stimuli used. These, together with the amount of work per cycle (in μJ), are indicated below each work loop. Reproduced with permission from Altringham and Johnston (1990a).

tested at 8 °C.) The force enhancement owing to stretch prior to shortening of active muscles generated maximum forces above isometric levels. The muscles of the 13 cm fish generated maximum power output at frequencies between 10 and 15 Hz, but the 67 cm fish performed maximally at 5 Hz. With a total strain of 10% of l_0 for fish of all sizes, strain rates at maximum power production varied between 1 and 1.5 $l_0\,s^{-1}$ for the 13 cm cod and was only about 0.5 $l_0\,s^{-1}$ for the fish of 67 cm. The length of the bundles of parallel fibres dissected from this size range of cod varied between 3.3 and 16 mm. Curtin and Woledge (1988) found a rectilinear correlation between fish length and fibre length in dogfish but, in contrast to the results obtained for cod, a constant shortening velocity at maximum power for a large range of body sizes at 12 °C. The optimal velocities, offering the largest power output, for cod during work loop experiments of the 13 cm fish were about twice as high as those of the 67 cm fish at 4 °C. This is consistent with tail beat frequencies used for swimming by fish of different sizes. With only one shortening velocity at maximum power for all sizes, most size classes of the dogfish would be operating suboptimally. This result is puzzling. The temperature difference could only be responsible for an overall difference in values. It is hard to imagine that dogfish muscles are so much less efficiently adapted to serve all sizes than those of cod.

Variation in temperature and size also plays an important role in yet another category of physiological experiments with lateral muscles of fish, not with single fibres or small fibre bundles but testing contractile properties of whole blocks of lateral muscle including several myotomes.

7.4 MAXIMUM MUSCLE TWITCH FREQUENCIES OF MYOTOME BLOCKS

The speed of a steadily swimming fish is the product of stride length and tail beat frequency (page 116). Each species seems to have a maximum stride length, and tail beat frequencies are generally used to modify speeds. For the highest speeds, a fish should swim using its best possible stride length at the highest tail beat frequency. One tail beat requires subsequent contraction of muscles on the left and right side of the body; therefore the minimum time to complete one tail beat is twice the minimum contraction time of the muscles on one side of the body (Wardle, 1975). Precise kinematic analysis is used to determine the stride length. Wardle developed a method to measure twitch contraction times using blocks of lateral muscles dissected from half-way down the body of freshly killed fish. A muscle block was rigidly fixed in a clamp at one end and attached at the other end by a hook to a displacement transducer, loaded with a light spring to give some isotonic tension. The muscle was submerged in a temperature-controlled saline solution. Single direct current pulses were applied to the muscle, either through the saline bath or via the stainless steel clamp. The resulting movement of the muscle was recorded. The time from the stimulating pulse to the peak of the contraction was measured and represented the shortest contraction time or twitch contraction time for that block of muscle at the given temperature. Experiments (Wardle, 1980) showed that even substantial variations in the muscle load had no influence on the twitch contraction times measured. This rather robust method was used to measure ranges of twitch contraction times of a variety of fish species, predicting the effect of size and temperature on the maximum swimming speeds. In predator–prey relationships absolute values of the highest speeds are of decisive importance for survival. Fig. 7.6 compiles data from various sources, offering a first impression of the temperature and size effects on the twitch contraction times. There is a steep

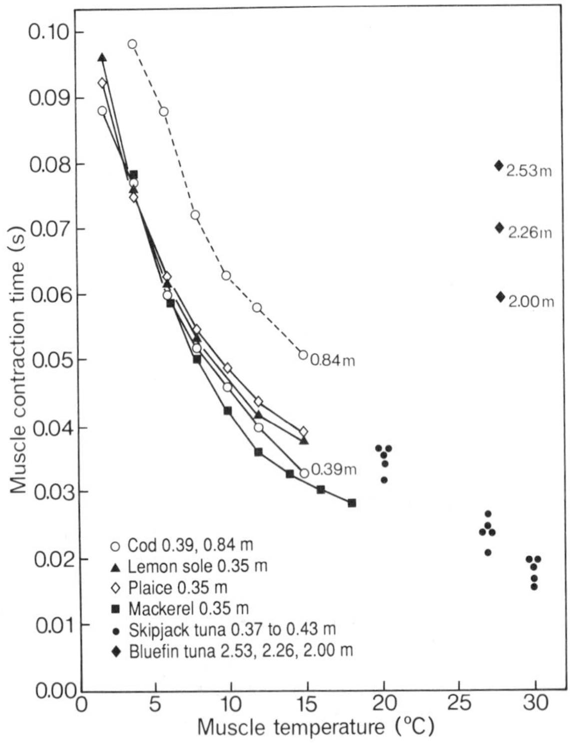

Fig. 7.6 Effect of temperature and body length on the twitch contraction time of white lateral muscle of the species and sizes indicated. Samples were taken approximately half-way down the body in each case. The skipjack tuna data are from Brill and Dizon (1979), those of the bluefin tuna from Wardle *et al.* (1989) and the others are from Wardle (1980).

decrease with increasing temperature, and this trend seems to be the same for species ranging from bottom-dwelling flatfish (plaice and sole, *Solea solea*) to purely pelagic swimmers (mackerel and skipjack tuna). The data points for two sizes of cod give an indication of the size effect: at the same temperature the smaller fish have shorter twitch contraction times. We shall deal in detail with temperature and size effects on maximum swimming speed in Section 7.5.

Estimates of maximum performance

Wardle's method can be used to analyse high-speed records (Wardle and Videler, 1980b) or to estimate maximum velocities. For example, Walters and Fierstine (1964) recorded a yellowfin tuna, *Thunnus albacores* of 98 cm moving at 20.8 $m\,s^{-1}$ (75 $km\,h^{-1}$) and a 113 cm wahoo, *Acanthocybium solandin*, at 21.4 $m\,s^{-1}$ (77 $km\,h^{-1}$). Both species are scombroids capable of stride lengths of about 1 body length; therefore the twitch contraction times of the muscles during these swimming records must have been 0.023 and 0.026 s respectively. These values appear extremely low if we compare them with the tuna data in Fig. 7.6. Even if the muscle temperatures were about 30 °C, we would expect values between 0.03 and 0.04 s for animals of that size. We can also use the data of Fig 7.6 to 'guesstimate' the maximum speed of what is probably the fastest fish in the world. Although proper direct evidence is lacking, I consider the swordfish as the most highly adapted to minimize drag (Section 4.4). Reduced drag probably enables this fish to make strides in the order of 1.2 body lengths. The estimate of 30 $m\,s^{-1}$ for a 3 m swordfish would in that case require a minimum contraction time of about 0.06 s, which seems not to be entirely unrealistic for an animal of that size, if it were swimming at temperatures as high as 30 °C.

Wardle *et al.* (1989) tried to estimate the maximum speed of the giant bluefin tuna, using a slightly different technique. The twitch contraction times of the anaerobic swimming muscles were not measured on blocks of muscle tissue but directly in the lateral muscle of pithed fish within 5 min after death. Bluefin tuna are caught during the summer on both sides of the Strait of Gibraltar in set nets as they leave the Mediterranean after spawning. The fish, varying in size between 1.7 and 3.3 m, are maintained and fed in cages where they gain weight before they are killed and sold to the Japanese market as *maguro no sashimi*, the most appreciated fresh fish meat in Japan. The fish were killed by a shot to the head from a 12-bore shot-gun. The head was severed through the region of the gills and a stainless steel wire was pushed along the length of the neural arch to destroy all spinal nerve reflexes. The measurements were made with a portable isotonic muscle twitch transducer, consisting of two stainless steel needles connected by a plastic arch (made of a rainwater pipe support clip) fitted with strain gauges.

The needles were pressed into the lateral muscle about 15 cm above the lateral line at various positions along the body. For each measurement a stimulation pulse of 2 ms was output to the needles, the muscle subsequently contracted and relaxed again. This movement was transduced through the strain gauges to a recorder. The time elapsed between the stimulus and the peak of the contraction was considered the twitch contraction time. Recordings were repeated at least three times at each spot along the body. Three fish, 2.00, 2.26 and 2.53 m, were measured. The temperature of the muscles at the twitch sites was 28 °C, which was 8 °C higher than the seawater temperature. Results, (Fig. 7.7) show a marked increase in twitch contraction time from head to tail. (I used the readings from half-way down the body in Fig. 7.6 to make the data points comparable with the other values.) Wardle (1985) previously reported a similar, nearly twofold increase of twitch contraction time down the body of mackerel, cod, haddock and saithe, so it seems to be a widespread if not universal phenomenon. It could change the stiffening of the rear part of the body during swimming at high speed. The thickest muscle in the front part of the body is shortening on alternate sides at a rate limited by the local twitch contraction time. Towards the tail, the contraction times of the opposing muscles will progressively overlap and stiffen the body. We will come back to this when considering force transmission during swimming (Section 8.5).

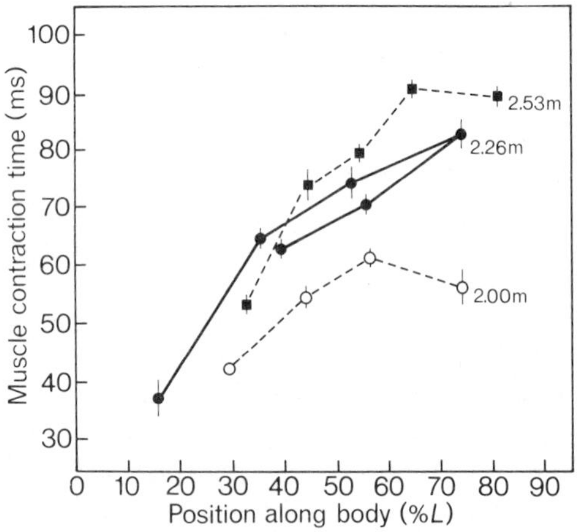

Fig. 7.7 Twitch contraction time (mean ± SD) of the white lateral muscle of bluefin tuna at different positions along the length of the body of three fish. The temperature at the twitch sites was 28 °C. The lines connecting the data points indicate the order in which the measurements were taken at the different positions. Measurements started at the front end in each case. Reproduced with permission from Wardle *et al.* (1989).

Stride lengths of the bluefin tuna were measured from video images of fish cruising in the cages measuring 80 × 20 × 20 m (length × width × depth). A remotely controlled video camera was mounted in fixed position on the bottom of the cage, looking straight upward. A range-measuring echosounder was attached to the camera. The echosounder gave an accurate measurement of the distance between camera and fish and made it possible, after calibration with rods of known lengths, to measure fish lengths and distances covered with an accuracy of ± 2%. Cruising speeds ranged from 0.6 to 1.2 $L\,s^{-1}$, with stride lengths varying between 0.54 and 0.93 *L* (the average being 0.65 *L*). A 2.5 m bluefin tuna using a twitch contraction time of its main bulk of muscle of 0.05 s and a stride length of 0.93 *L* would swim with a tail beat frequency of 10 Hz at 23.25 $m\,s^{-1}$ (83.7 $km\,h^{-1}$). This is obviously an over-estimate: a conservative estimate for the same fish would be 9.75 $m\,s^{-1}$ (35 $km\,h^{-1}$), with a stride length of 0.65 *L* and a tail beat frequency of 6 Hz.

In conclusion: estimates of maximum speeds of the fastest fishes desperately need confirmation by actual measurements.

7.5 SIZE AND TEMPERATURE EFFECTS ON CONTRACTION VELOCITY

The previous section made it clear that both temperature and body size have a marked effect on the maximum muscle twitch frequencies and therefore on the maximum speed at which a fish can swim. The questions remain, how big these effects are and whether they are consistent for different species. Videler and Wardle (1991) reported on the analysis of a large data set of twitch frequency measurements of cod, varying in size between 20 and 85 cm, at seven different temperatures between 2 and 15 °C, to quantify the relations. The rules of thumb found with this data set are compared below with the data available from the literature to establish their generality.

Temperature effect on maximum performance

Physiological effects of temperature change are generally described as $Q_{10\,°C}$ effects, where $Q_{10\,°C}$ represents the increase in a rate caused by an increase in temperature of 10 °C. If a rate doubles over a 10 °C increase, $Q_{10\,°C}$ is 2; if it triples, $Q_{10\,°C}$ is 3; and so on. If R_1 and R_2 are the twitch frequencies at two temperatures $t1$ and $t2$,

$$R_2 = R_1 \,.\, Q_{10\,°C}{}^{(t2-t1)/10} \tag{7.1}$$

The $Q_{10\,°C}$ is calculated from the equation:

$$Q_{10\,°C} = (R_2/R_1)^{10/(t2-t1)} \tag{7.2}$$

Table 7.1 Twich contraction frequencies (Hz), $Q_{10^\circ C}$ and $Q_{10\,cm}$ values for cod, *Gadus morhua*, of different body lengths, measured at a range of temperatures. The mean value of $Q_{10^\circ C}$ is 2.06 and that of Q_{10cm} is 0.89

Length (cm)	*Temperature (°C)*							$Q_{10^\circ C}$
	2	*4*	*6*	*8*	*10*	*12*	*15*	
20	18	26	26	30	38	42	46	2.00
39	12	14	16	20	22	24	28	1.95
52	10	14	16	18	20	22	26	1.91
53	10	12	16	18	22	26	30	2.24
65	8	10	12	14	16	18	22	2.11
67	8	10	12	14	18	20	22	2.13
77	10	10	12	14	18	20	24	2.13
85	8	10	12	14	16	18	20	1.97
$Q_{10\,cm}$	0.89	0.88	0.89	0.89	0.88	0.89	0.89	

In practice it is calculated from regressions of the logarithms of twitch frequencies against temperature or by non-linear curve fitting through the original data. Handbooks on physiology provide more background information on $Q_{10\,^\circ C}$ effects in general (e.g. Schmidt-Nielsen, 1990). We are not only interested in the effect of temperature on muscle twitch frequencies, but also in the impact of differences in body length.

Table 7.1 shows the twitch contraction frequencies in Hz at seven temperatures between 2 and 15 °C, obtained from cod of eight sizes, varying from 20 to 84.5 cm. It is part of the data set on which Wardle based his 1975 and 1977 papers on size and temperature effects (see also Fig. 7.6). The mean $Q_{10\,^\circ C}$ value for the eight length classes calculated from regressions of log frequency against temperature is $Q_{10\,^\circ C} = 2.06$ (SD 0.1, $N = 8$). This result predicts that the maximum stride frequency and hence the speed of a cod doubles with every 10 degree temperature increase. Langfeld *et al.* (1989) found a $Q_{10\,^\circ C}$ value of 2 for the time to build up maximum tension in isolated bundles of fast fibres of the bullrout, between 0 and 16 °C. A $Q_{10\,^\circ C}$ of 2 was also found for amphibian muscles by Edman *et al.* (1976) in *Rana*, and by Lannergren *et al.* (1982) in *Xenopus*. $Q_{10\,^\circ C}$ values of about 2 are common for the rate of speeds of biological enzyme-catalysed chemical reactions. The tail beat frequencies of newly hatched herring and plaice larvae show a $Q_{10\,^\circ C}$ of 1.9 at temperatures between 5 °C and 15 °C (Batty and Blaxter, 1992).

Length effects on maximum performance

Analogously to the more familiar $Q_{10\,^\circ C}$, a Q_{10} value for every 10 cm difference in length, the $Q_{10\,cm}$, can be defined as the ratio of the muscle twitch frequencies for each 10 cm difference in length. The Q_{10cm} values for cod muscle obtained from 20–84.5 cm fish are similar at the different temperatures (Table 7.1). The average value is $Q_{10\,cm} = 0.886$ (SD 0.006, $N = 7$). We could now predict the maximum tail beat frequency for cod ($TBF_{max\ cod}$), using $Q_{10°C} = 2.06$, $Q_{10\,cm} = 0.886$ and the tail beat frequency of 15.32 Hz of a 20 cm cod at 10 °C as a bench mark, thus:

$$TBF_{max\ cod} = 15.32[0.886^{(L-20)/10}(2.057^{(t-10)/10})] \qquad (7.3)$$

where L is the body length in cm and t denotes temperature in °C.

Twitch contraction times at 14 °C for 58 samples taken from 26 Atlantic salmon of different lengths, ranging from 6 to 56 cm, yielded a $Q_{10\,cm}$ value of 0.876 (C.S. Wardle's unpublished results).

Table 7.2 lists the $Q_{10\,cm}$ values calculated from data in the literature. Archer *et al.* (1990), used nerve–muscle preparations of white fibres from the abdominal region of cod to measure twitch frequencies. Calculation of the $Q_{10\,cm}$ value over their total size range yielded 0.89. Curtin and Woledge (1988) made, as we saw earlier in this chapter, force–velocity measurements on intact fibres over a large size range of dogfish. They obtained a $Q_{10\,cm}$ value of 0.9.

Altringham and Johnston (1990b) calculated the length effect on optimal contraction frequencies of cod varying in size between 13 and 67 cm at 4 °C. Their results yield a $Q_{10\,cm}$ value of 0.84.

Fig. 7.8 shows a nomogram of the combined effects of temperature and length on the maximum swimming speed of cod based on the $Q_{10°C}$ and $Q_{10\,cm}$ values used in Equation 7.3.

Table 7.2 Values of $Q_{10\,cm}$ compiled from the literature

Species	*Temp.* (°C)	*Length range* (m)	*Max. TBF range* (Hz)	$Q_{10\,cm}$	*Source**
Gadus morhua	12	0.09–0.45	20.5–12.8	0.89	1
Salmo salar	14	0.06–0.56	20.1–10.4	0.88	2
Girella tricuspidata	14	0.14–0.45	20.2–12.4	0.85	3
	10	0.14–0.45	13.4–9.5	0.89	3
	6	0.14–0.45	9.4–5.7	0.85	3
Scyliorhinus canicula	12	0.15–0.65	21.5–13.0	0.90	4

*Sources: 1, Archer *et al.* (1990); 2, C.S. Wardle (unpublished results); 3, McVean and Montgomery (1987); 4, Curtin and Woledge (1988).

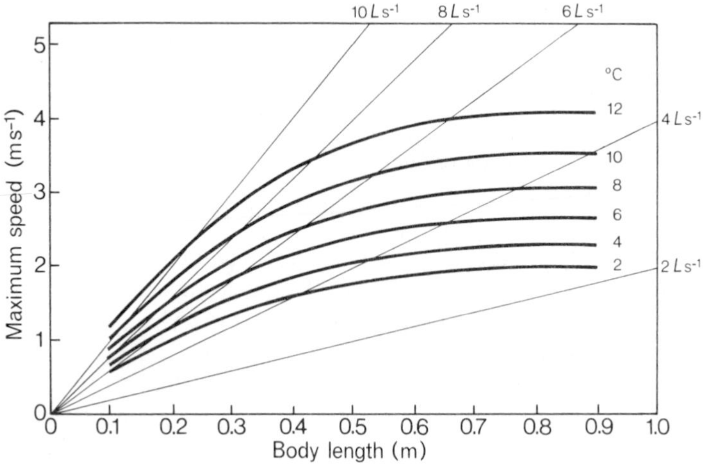

Fig. 7.8 Effect of length and muscle temperature on the maximum swimming speed of cod. The stride length is assumed to be 0.6 *L*. The thick curves, calculated using Equation (7.3), show the maximum speed in $m\,s^{-1}$ for the range of body lengths represented on the *x*-axis. The thin lines indicate velocities in $L\,s^{-1}$. Based on Videler and Wardle (1991).

Swimming at different temperatures

So far we have looked at the effect of temperature on the maximum performance of fish. Rome (1990) took a different viewpoint by looking at muscle performance at a given speed for different temperatures. He pointed out that the mechanics of locomotion for a given speed are independent of temperature. Swimming at a certain speed requires a certain stride length combined with the appropriate tail beat frequency. This implies that sarcomere length excursions and muscle shortening velocities are exactly the same at all temperatures. The mechanical power needed to swim at a certain speed is roughly proportional to the cube of the velocity and does not depend on the temperature. The force, and hence the mechanical power per unit cross-sectional area, that a muscle generates at a given shortening speed decreases at lower temperatures. Muscles use less energy at lower temperatures but the efficiency, defined as the rate of energy utilization per unit mechanical power produced, decreases substantially. Fish compensate for the diminishing power output of their muscles at low temperatures by recruiting more muscle fibres and faster fibres at a given speed, and by swimming at lower maximum speeds. Efficiency of muscle is maximal at velocities between 0.2 and 0.4 V_{max} (Fig. 7.3 (b)); it drops to zero both at V_{max} and when it operates at P_{max}. Rome *et al.* (1990) found that at each temperature, carp uses its lateral muscles within the range of most efficient ratios of V/V_{max}. The fish starts to recruit its white muscles when the V/V_{max} of the red ones becomes too high. When, during steady swimming, the V/V_{max} becomes too low, carp raises its contraction speeds by adopting the energy-saving burst-and-coast swimming pattern. At lower temperatures white muscles are recruited at lower swimming speeds, a strategy which decreases the maximum sustainable speed of the fish. The lowest steady swimming speed of carp at 10 °C is about 1.2 $L\,s^{-1}$, and at 20 °C it reaches about 1.6 $L\,s^{-1}$. A fish swimming at 1.2 $L\,s^{-1}$ at 20 °C will always use a burst-and-coast swimming mode. Intermittent swimming at high velocities will have a similar effect on the efficient use of white muscle.

Fish are usually very sensitive to sudden changes in temperature. On the other hand, temperate freshwater species tolerate large temperature differences between summer and winter when there is plenty of time for acclimation. The effects of acute and seasonal temperature changes on muscles performing oscillatory work have been studied by Johnson and Johnston (1991) who worked on white muscle fibres isolated from the abdominal myotomes of the bullrout, a marine species. The work loop cycles were optimal at strains of ± 5% of the resting length, independent of temperature. The number and timing of stimuli were adjusted to optimize net positive work output over a range of frequencies as explained above. Fish

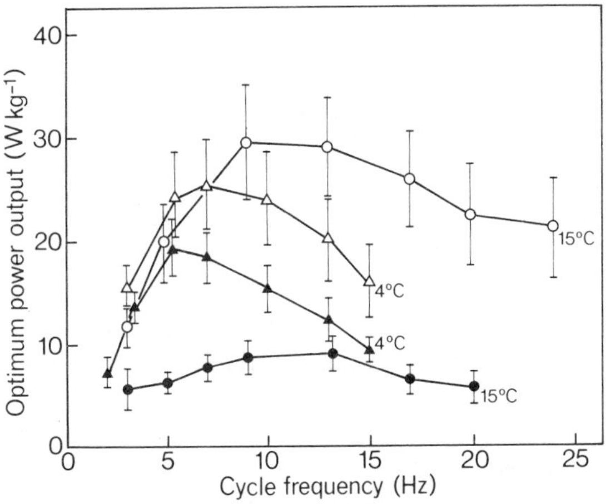

Fig. 7.9 Relationship between cycle frequency of white muscle fibre bundles from the abdominal myotomes of the bullrout, *Myoxocephalus scorpius*. The open symbols represent summer- and the closed symbols winter-acclimatized fish. The strain amplitude during the sinusoidal changes in length was ± 5% of the resting length. Cycle frequency and stimulation parameters were adjusted to maximize work output. The muscles were tested at 4 and 15 °C. Reproduced with permission from Johnson and Johnston (1991).

of similar size were caught during the winter and the summer. They were kept at ambient temperatures in each case; these were 4–5 °C in the winter and 12–13 °C in the summer. Fibres from summer fish and winter fish were tested at 4 and 15 °C. The cycle frequency required for maximum power output was around 5–7 Hz at 4 °C and between 9 and 13 Hz at 15 °C for both summer and winter fish (Fig. 7.9). The maximum power output of the summer-acclimatized fish was only slightly less when tested at 4 °C than at 15 °C. The winter-acclimatized fish had a maximum optimal power output of about 20 W kg^{-1} at 4 °C and less than 10 W kg^{-1} at 15 °C. The difference between the performance of both groups when tested at 15 °C is striking. The muscles of the fish in summer condition generate 30 W kg^{-1}, and the fibres of the cold-adapted fish only about 9 W kg^{-1}. The experiments indicate that these fish are better able to swim at different temperatures during the summer than during the winter, when they would have difficulties in coping with high temperatures.

Exactly the opposite effect is shown to exist in cyprinids. Swimming performance of carp, goldfish and roach is improved at low temperatures following a period of cold acclimation. Both V_{max} and P_{max} increased. Cold acclimation changed the contractile properties of the muscles by inducing changes in the molecular structure of myosin in the thick filaments (Johnston *et al.*, 1990). Freshwater fish in temperate zones experience temperatures close to 0 °C during winter periods. The muscles of cyprinids seem to be well adapted to such extremes.

7.6 ELECTROMYOGRAPHY

So far we have mainly been looking at functions of muscle *in vitro*. The technique on which we now focus studies muscle function in live fish. Electromyograms (EMGs) are recordings of electrical activities in the active muscle. The nature of these electrical activities is not at all clear, nor is the relation between the signal and the timing and magnitude of the force exerted by the muscle. Fig. 7.10 shows Wardle's (1985) recording of the force development, the muscle action potential picked up by the EMG electrode and the stimulus pulse given to a block of lateral muscle of cod. Note how the EMG signal starts simultaneously with the stimulus pulse, but ends well before the maximum force is developed. This implies that the start of an EMG indicates the start of muscle activity but that an EMG does not provide information about the timing of the peak force.

Traces from EMG electrodes in the red muscles of free-swimming mackerel (from He, 1986) nicely demonstrate the alternating activities of muscles on both sides of the body (Fig. 7.11 (a)) and the timing of the onset of activity between two locations on one side of the body, one behind the other. These

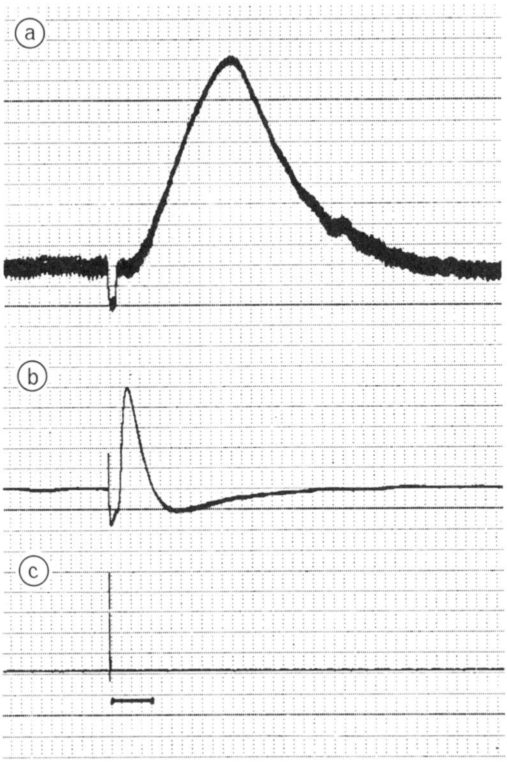

Fig. 7.10 Recorder traces showing the temporal separation of the electrical and mechanical activities in a block of lateral swimming muscle of a cod (length 30 cm, tested at 14 °C). (a) Record of muscle force, rising to a peak at 30 ms. (b) The electrical activity in the muscle measured with an EMG electrode. (c) The stimulus pulse of 10 μs. The horizontal bar is 10 ms on the time scale. Reproduced with permission from Wardle (1985).

traces (Fig. 7.11 (b)) show a great deal of overlap, in that the one closest to the head starts before the one near the caudal peduncle and lasts longer.

EMGs can be used to establish patterns of rhythmic activity of muscles. I tried to do that with the nine intrinsic muscles in the tail of tilapia (Videler, 1975), using up to six very small bipolar electrodes simultaneously (Videler and Beukema, 1973). The ultimate goal that my PhD supervisor, the ethologist G.P. Baerends, and I were trying to reach was to design a means of scoring muscle contractions, analogous to musical notation, which would represent a certain behaviour. We dreamt of an anaesthetized fish with electrodes in its muscles, connected to a keyboard. If we knew the score to play we could in our fantasy make it perform behaviour as complex as

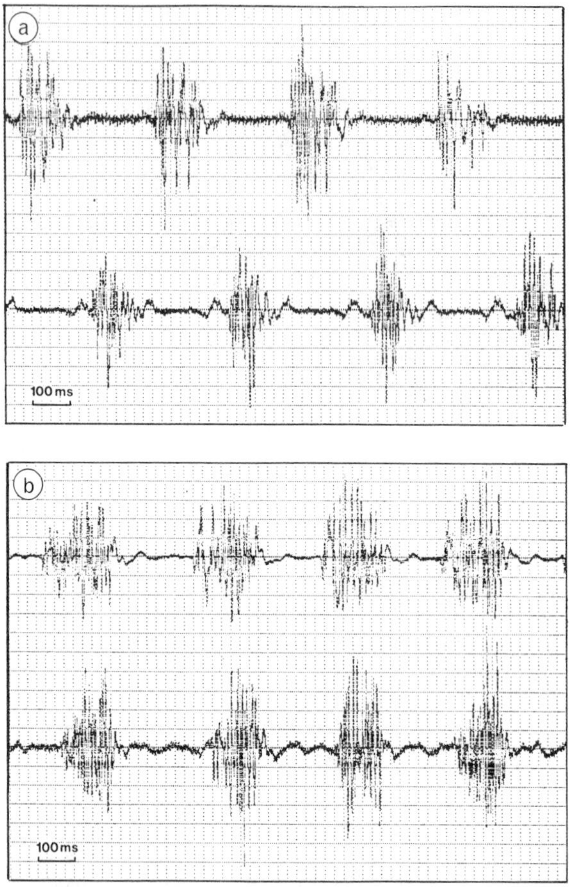

Fig. 7.11 Traces of EMG recordings from the red muscle of free-swimming 0.3 m long Atlantic mackerel, *Scomber scombrus*. (a) The electrodes were positioned on opposite sides of the fish at a distance of about 0.75 of the body length (L) from the nose. The upper trace is from the right- and the lower from the left-hand side of the fish. (b) EMG recordings from electrodes positioned on the same (right hand) side of the body. The upper-trace electrode is the most anterior one at about 0.37 L from the nose. The electrode giving the lower trace is at 0.60 L. Reproduced with permission from He (1986).

courtship display. I managed to record courtship display from a male tilapia carrying six electrodes in its tail muscles, but never attempted the inverse procedure. The reason was that even during steady, straight forward tail beat cycles, not a single movement was an exact replica of the next, if the activities of six muscles were observed. There was an overall pattern, in that for instance during a tail stroke to the right the main bulk of muscles on the right-hand side were active, but also there were always muscles on the opposite side that showed some activity.

Electromyography has been a useful method to study the activity of red and white lateral muscles in fish during steady locomotion at various speeds and an appropriate tool to detect the use of muscles during startle responses. I will refer to this body of knowledge in the next chapter by summing up and relating the existing morphometric, kinematic, dynamic and electromyographic knowledge to achieve a comprehensive view of how straight forward fish swimming movements with body and tail are made.

7.7 SUMMARY AND CONCLUSIONS

Elaborate physiological techniques disclosed various mechanical properties of lateral fish muscles and showed in each case, where this could be determined, that the muscles are extremely well adapted to the task they must perform. Red and white carp muscles operate near their optimum sarcomere length values. The muscle fibres of dogfish and carp generate maximum power at velocities between 0.2 and 0.4 times their maximum velocity, and that is exactly where they usually operate, the red fibres at lower speeds than the white ones. The white fibres make use of their helical orientation to shorten over an equal distance. Their position closer to the vertebral column allows them to contract over a shorter distance than the red fibres, given the same body curvature. This shorter distance decreases the contraction velocity during the fast movements which they power, and brings it into the range where they can generate maximum power.

Lateral muscle fibres perform optimally during cyclic activities when they are stimulated during lengthening, just prior to the shortening period of the cycle. Length changes of $\pm$ 3–5% of the resting length give maximum work loops. Red muscles at low cycle frequencies generate power in the order of 6 $W\,kg^{-1}$, and white fibres at high frequencies about 30 $W\,kg^{-1}$.

Lateral muscle twitch frequencies are postulated to determine the maximum velocity at which a fish can swim given a certain maximum stride length. Twitch frequencies depend on temperature with a $Q_{10\,°C}$ of about 2 and on body length with a $Q_{10\,cm}$ of 0.89.

Fish are, within limits, able to recruit enough muscle to maintain a given speed at a certain temperature. At a lower temperature the power of the muscles decreases and more fibres are recruited to maintain the same swimming velocity.

The muscles of cold-adapted fish perform badly in terms of power production at higher temperatures. Fish acclimatized to warm temperatures generate a high power output at both high and low temperatures. The power output at low temperatures is even higher than that of cold-acclimatized fish.

Electromyographic studies can be used to show patterns of muscle activities during swimming. Patterns between muscles on the left and right side of the body and between muscles at different positions between head and tail will be discussed in the next chapter. These patterns provide crucial information needed to understand how the lateral muscles propel the fish.

Chapter eight

Swimming dynamics: exchange of forces between fish and water

8.1 INTRODUCTION

Knowledge of swimming movements, hydrodynamics, anatomy and physiology needs to be combined to assemble a coherent picture of undulatory propulsion in fish. Unfortunately, the available data originate from a variety of species and are difficult to compare because of the different methods that have been used to obtain them. The method for kinematic analysis of steady swimming with cyclic undulations of body and tail, described on pages 108–12, will be used to compare the movements of saithe and eel. This method allows us to apply a dynamic analysis which uses detailed knowledge of the movements not only to predict forces and bending moments inside the fish but also to estimate the reactive forces exerted by the water on the fish body from head to tail. The forces and bending moments are related to the waves of curvature along the body, and the results show different phase relationships between the two for saithe and eel. The results predict the timing of the peak forces in the muscles.

Electromyographic techniques are the most promising to show the timing of muscle activity in a swimming fish. The amount of published data is limited and the connection with the dynamic study predictions is not immediately obvious. Electromyographic data from free-swimming saithe and mackerel are used to link the results from the kinematic and dynamic approach with muscle function. The timing of activation, superimposed on the cyclic changes in length of the muscle fibres, alters along the body from head to tail. The main power is generated in the first half of the fish and the largest forces are produced by the fibres in the caudal peduncle. Several possibilities for the transfer of power from the anterior part of the fish to the tail, where the power is spent on the water, are discussed. Muscles are usually expected to generate forces in the fibre direction, either by shortening against a load or by resisting stretch caused by an external force. Force generation by bulging is an alternative and fish muscles are probably using it.

8.2 DYNAMICS OF INTERACTIONS BETWEEN UNDULATING FISH AND WATER

Kinematic analysis of simple, straight, forward swimming at close to uniform speed will be the basis of the following quantitative study of swimming dynamics. Chapter 5 ended with a kinematic analysis of saithe swimming with periodic lateral oscillations of the body, showing that the lateral displacements and the changes in body curvature behave as simple harmonic motions. Kinematic data of steadily swimming saithe and eel, obtained with the same method, will be used to demonstrate some of the principles involved in the interaction between bending forces and bending moments in the body and the reactive forces from the water.

Kinematics of steadily swimming saithe and eel compared

Table 8.1 offers the primary and some derived kinematic quantities for the two examples. Examples of the movements of saithe are shown as digitized outlines from film frames and computed centre lines in Fig. 5.3. The fish are

Table 8.1 Kinematic and morphometric quantities for saithe and eel

Variable	*Unit*	*Symbol*	*Saithe*	*Eel*
Length	m	L	0.37	0.14
Period	s	T	0.278	0.276
Speed	m s^{-1}	u	1.14	0.28
Relative speed	$L\ s^{-1}$	U	3.09	2.0
Stride length	$L\ T^{-1}$		0.86	0.55
Maximum tail amplitude	L	A	0.083	0.102
Relative wave speed	$L\ s^{-1}$	V	3.74	2.86
		U/V	0.82	0.69
Tail height	L		0.24	0.096
Body volume	L^3		0.0113	0.0018
Wetted surface area	L^2		0.401	0.196

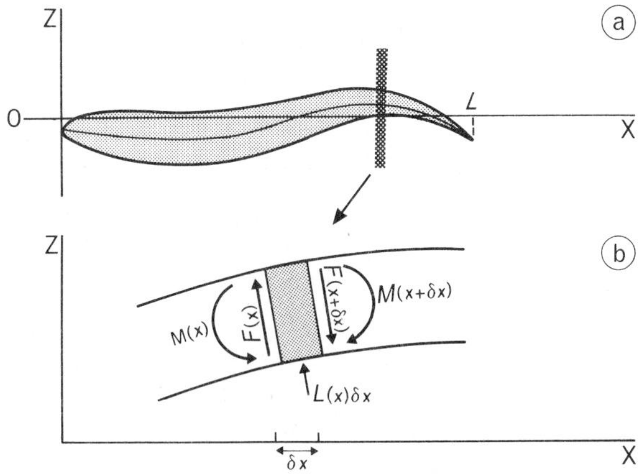

Fig. 8.1 (a) Schematic dorsal view and centre line of a fish, indicating the co-ordinate system used. The tip of the nose is at $x = 0$ and the tail tip at $x = L$. A thin slab of fish is enlarged and shown in (b). (b) Forces and bending moments acting on a thin slab generated by the fish and the reactive force from the water acting on a thin slice of a steadily swimming fish body. For explanation see text. Redrawn from Hess and Videler (1984).

swimming in an X–Z coordinate system drawn in Fig. 8.1. (Y is the vertical axis in this system). The nose of the fish is at $x = 0$ and the tail end at $x = 1$; the horizontal unit is the body length L. The fish is swimming to the left and the water flows relative to the fish in the X-direction at the swimming speed. The z-axis points laterally. The centre line of the fish is described by:

$$z = h(x,t),\ 0 \leqslant x \leqslant L \tag{8.1}$$

where h is the lateral displacemement of the centre line.

Both fish were swimming steadily at uniform cruising speeds. The swimming styles are different. The maximum tail tip amplitude of the eel is about 25% higher than that of the saithe (amplitudes are relative to body length L). The amplitude of the maximum lateral deflection of the body from head to tail increases almost linearly to the rear for the eel. The head of saithe has a larger amplitude than the eel's and it decreases rearward to a minimum value just behind the head; from there the amplitude increases rapidly to the highest value at the tail tip (Fig. 8.2(a)). The right-hand side of that figure shows that the maximum lateral deflection of saithe uses 1 period T to go from head to tail whereas that of the eel is slower and takes 1.3 T to travel along the body. There are also substantial differences in the way the curvature (expressed as the inverse of the radius of curvature) travels down the body (Fig. 8.2(b)). The maximum curvature of the eel is much larger at every point of the body. The curvature is small in the anterior part of saithe and

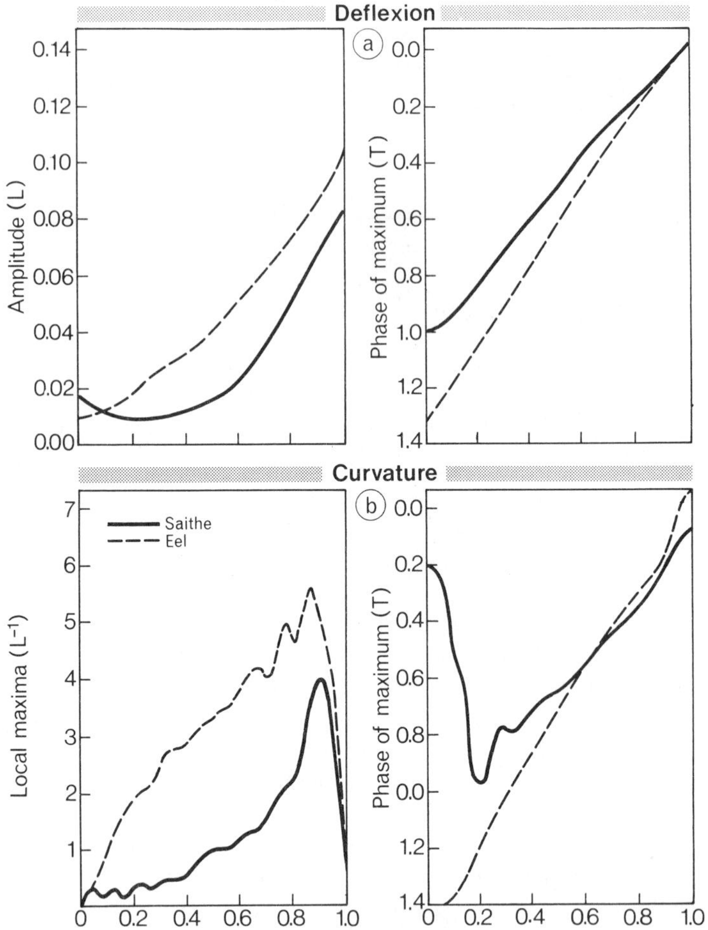

Fig. 8.2 (a) Lateral deflection $h(x,t)$ along the body of steadily swimming saithe (solid curves) and eel (broken curves). The horizontal unit is body length L. Left, amplitude of the deflection; right, phase of the maximum lateral deflection as a function of the position on the body between head and tail. The unit of time is the period T. The instants of maximum deflections of the tail tips were arbitrarily chosen to be at $T = 0$. (b) Lateral curvature along the body in saithe (solid curves) and eel (broken curves). Left, local maxima expressed in units L^{-1}; right, phase (unit T) of the maximum curvature as a function of the position on the body. Small-scale undulations are due to noise in the original data. The first part of the curves for saithe ($x < 0.2$) have no physical meaning because the head of saithe is large and stiff and does not show any curvature at all. Redrawn from Hess (1983).

reaches a maximum at the tail (it is considered to be 0 at the tail tip by definition). It is more evenly distributed along the eel. The velocity of the local maxima is again faster in saithe than in eel.

Dynamic analysis of saithe and eel swimming

The fish are treated as self-bending thin flexible rods under the influence of hydrodynamic forces. The approach is visualized in Fig. 8.1(b). Forces and moments are acting on an arbitrary thin slice of fish perpendicular to the backbone. The dimension of the slice in the longitudinal direction of the fish is δx. The anterior part of the body exerts a force $F(x)$ on the posterior part. The moment exerted by the anterior part attempting to turn the posterior part counter-clockwise is $M(x)$. If we assume δx to be infinitely small, the force from the posterior part on the anterior part is $-F(x)$ and the clockwise bending moment $-M(x)$. The lateral force exerted by the water on the slice per unit length is $L(x)$. It represents the rate of change of momentum of the water near that slice, proportional to mass times acceleration of that water. The total net force acting on the slice must equal the slice's mass times its lateral acceleration. If the fish's body mass per unit length is $m_b(x)$, we obtain the equilibrium of forces:

$$[-\delta F(x,t) \;/\; \delta x] + L(x,t) = m_b(x)\; \delta^2 h(x,t) \;/\; \delta t^2 \qquad (8.2)$$

The net bending moment, $\delta M(x,t) / \delta x$, acting on the slice is approximately equal to $-F(x,t)$. Substitution in (8.2) yields:

$$\delta^2 M(x,t) \;/\; \delta x^2 = [m_b(x)\; \delta^2 h(x,t) \;/\; \delta t^2] - L(x,t) \qquad (8.3)$$

Bending moments on the swimming fish can be calculated by integrating the left-hand side of this equation twice, but the right-hand side has to be solved first. Measurements of the distribution of the body mass of the fish between head and tail were made by cutting a deep-frozen fish into thin slices with a fine diamond saw. The results of the weighing exercise are shown in Fig. 8.3. The lateral acceleration of every part of the body as a function of time is provided by the kinematic analysis. The hydrodynamic lateral force per unit length, $L(x,t)$, was determined with Lighthill's (1960) slender body theory. Application of this theory requires knowledge of the kinematics of each part of the fish and an estimate of the virtual mass of water, $m_a(x)$ accelerated by each slice of the fish (see pages 16–20 for an explanation of the principles of this approach). So far the method seems to be rather straightforward, but in reality, several physical and mathematical problems had to be solved. For example, theory requires that bending moments vanish both at the point of the head ($x = 0$) and at the tip of the tail ($x = 1$). One would expect this to happen automatically but that is not the case: small errors in the parameters that could be measured and deviations between a real fish and Lighthill's

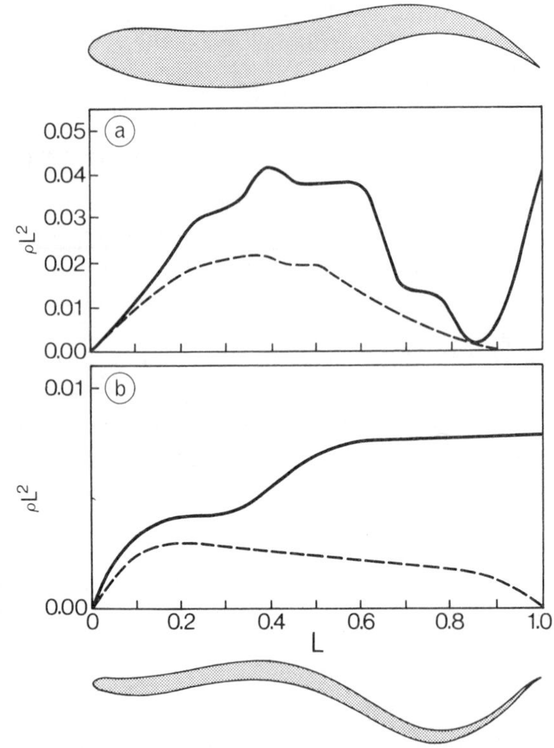

Fig. 8.3 Distribution of body mass $m_b(x)$ (broken curves) and virtual mass of water $m_a(x)$ (solid curves) for saithe (a) and eel (b). The unit of the x-axis is L and of the y-axis ρL^2. Note the fivefold y-scale difference. Redrawn from Hess (1983).

slender body made corrections necessary. (The eel is closer to a real slender body because it needed much smaller corrections than saithe.) This problem was solved by adding a certain amount of stiff yawing (recoil) motion ($A(t) + B(t)x$) to the lateral deflections $h(x,t)$. Further details of the mathematical model used are beyond the scope of this book, and the interested reader is referred to the section describing that model in Hess and Videler (1984). At this stage it should be mentioned that the large number of uncertain factors and estimated values make this theoretical approach rather crude. Having said that, we can turn to the results for saithe and eel.

Calculated maximum bending moments from head to tail for the two example cases are shown in Fig. 8.4. The curves on the left reveal that the maximum bending moments and the maximum power per unit fish length is found around 0.6 L in both species. The solid curves on the right give an impression of the velocity of the instant of maximum bending moment when it travels from front to rear compared with the speed of the

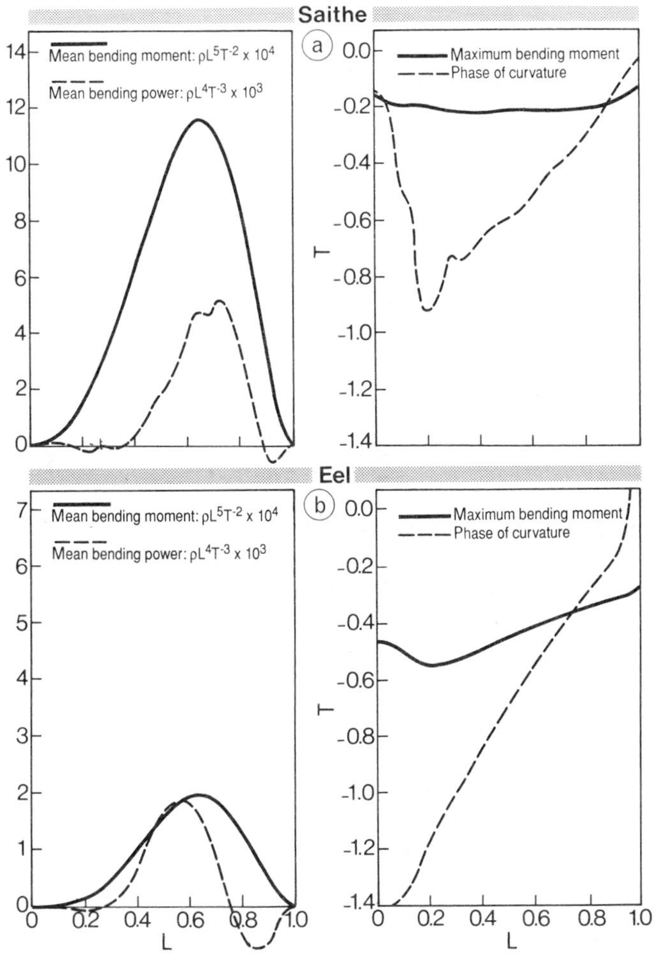

Fig. 8.4 (a) Solid curves: bending moment $M(x,t)$ for saithe. Left: the maximum values for each x-position from head to tail. The dimensionless unit is $\rho L^5 T^{-2} \times 10^4$, in conventional terms 91.97×10^4 N m. Right: the phase of the maximum bending moment as a function of the position on the body compared with (broken curve) the phase of the curvature shown also in Fig. 8.2(b) (the unit is T). Broken curve left: saithe, the mean bending power per unit length, where the unit is $\rho L^4 T^{-3} \times 10^3$ (dimensionless) and 89.41×10^4 W m^{-1} (conventional). (b) The same set of curves for the eel. The conventional unit for the bending moment is 72.37×10^2 N m and for the mean bending power 18.73×10^3 W m^{-1}. Redrawn from Hess (1983).

wave of maximum curvature. In each species these events are out of phase. The result for saithe is most surprising because it shows an almost instantaneous occurrence of the maximum bending moment over the whole body length at about 0.2 T before the maximum curvature reaches the tail tip. The velocity of the maximum curvature is much slower because it takes about 1 period T to run from head to tail. The instant of maximum curvature of the eel is slower by using about 1.4 T to cover the length of the body. The maximum bending moment covers that distance in about 0.3 T and arrives at the tail tip 0.3 T before the maximum curvature after starting 0.8 T later near the head. The conclusion is that the wave of the maximum bending moment generated by the muscles (*WoB*) runs from head to tail at a much higher speed than the wave of maximum body curvature (*WoC*). This disparity has important implications for the relation between lateral muscle function and swimming movements, the subject of the next section.

8.3 MUSCLE ACTIVITY FROM HEAD TO TAIL

To investigate the consequences of the disharmony between *WoB* and *WoC* , I replotted the results for saithe and eel, as shown by the graphs on the right-hand side of Fig. 8.4, in a simplified way in Fig. 8.5. The horizontal axis spans, once again, the length of the fish from head to tail; vertically the time is plotted using the period T as the unit of time. The time runs from the bottom to top of the graphs. The *WoB*s are plotted as thick lines, the bending moment to the right is a broken line, the one to the left solid. The *WoC*s are represented by the thin lines, broken to the right and solid to the left. The *WoB*s of the eel take slightly more than 0.3 T to run down the body, which is about 3.8 times faster than the speed V of the *WoC*s. The *WoB*s of saithe are more or less instantaneous and the *WoC*s run at approximately 1 LT^{-1}. Let us first consider saithe. From behind the head up to about 0.4 L the maximum bending moment occurs on both sides after the instant of maximum curvature (the bending curve lies above the line indicating the curvature on the same side). This implies that the muscles are stretching at the instant the bending moment is maximal. In other words, in that region of the body the muscles generate force when they are being stretched (excentric contraction). At 0.4 L the *WoB* passes at the instant of maximum curvature. Between 0.4 L and 0.85 L, maximum curvature occurs after the instant of maximum bending, which means that muscles are shortening when the maximum bending moment happens. The highest bending moment and the highest values of power per unit length are found in this region of the body. The position 0.9 L is on the caudal peduncle where the maximum bending moment on one side coincides with maximum curvature on the other side. The use of muscles in the swimming eel is different. From

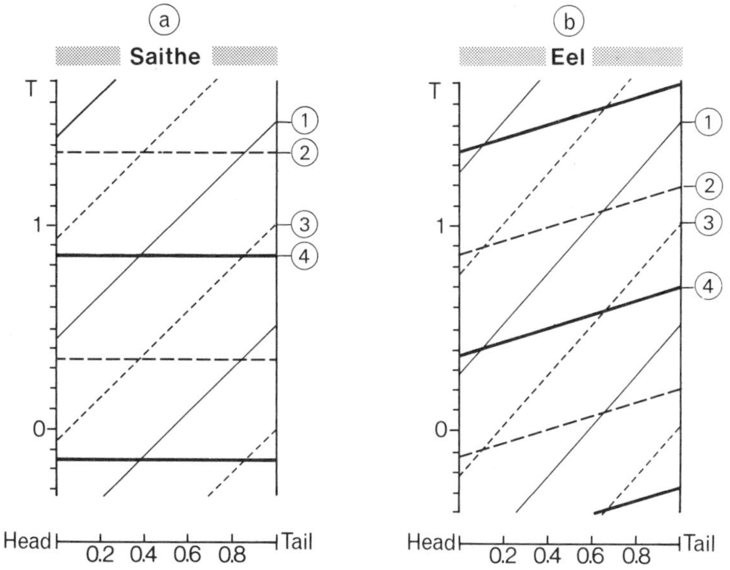

Fig. 8.5 Phase of maximum bending moments and maximum curvature of saithe (a) and eel (b). The tail beat period T is the unit of time. Speed of the wave of curvature of saithe is approximately 1 LT^{-1}, that of the eel 0.8 LT^{-1}. The bending moment of saithe is instantaneous from head to tail. The bending wave of the eel runs at 3 LT^{-1} along the body. The instant $T = 1$ coincides with the leftmost tail tip position, when the maximum curvature to the right just ended at the tail tip. 1, Maximum curvature to the left; 2, maximum bending moment to the right; 3, maximum curvature to the right; 4, maximum bending moment to the left. Based on Videler (1985).

behind the head at 0.1 L to about 0.7 L the maximum bending moment is exerted while the muscles are shortening. Between 0.7 L and the tail, lateral muscles are active during a stretching phase. Muscle activity patterns should match these findings based on kinematic analysis of swimming fish and a mathematical model approach. But EMGs do not indicate the instant of maximum force produced (Fig. 7.9), they give a good indication of the instant of onset of muscle activity. If a fixed relation were to exist between the timing of the onset of muscle activity and the instant of maximum force, we would expect the two to run down the body at the same velocity. However, as long as this is not clear, the onset of EMG activity is treated as a separate phenomenon. In summary, we now have to deal with three waves running down a swimming body, each with its own specific velocity: the *WoC*, the *WoB* and the wave of onset of muscle activity (*WoMO*).

Several myographic studies of lateral muscles in swimming fish provide evidence for the existence of disharmony between the travelling speeds of the lateral undulations and onset of muscle activities.

Electromyographic evidence

Such evidence of disharmony already existed long before the dynamic studies on saithe and eel swimming were published. Blight (1976) made EMGs of the lateral muscles of larval newts and of the tench during swimming and linked these to simultaneously filmed swimming movements. Two recordings were published where the newts showed simultaneous activity of the anterior and posterior electrodes on the same side during undulatory swimming. In one recording, the start of the signal of the posterior electrode consistently followed that of the anterior one, showing a small delay. The muscles of the tench, measured with three electrodes on each side, were sequentially active. Within the resolution of 10 frames per period T, the contractions of the myotomes matched the observed waves of bending on the fish reasonably well. Blight, however, did point out that there is a brief period in the swimming cycle when there is EMG activity simultaneously in all three electrodes on one side. The muscles at the head-end electrode are terminating their electrical activity just as those near the tail are starting. Blight (1976, 1977) reached the conclusion that *WoC*s running down the body of swimming animals might be generated by alternating instantaneous activity of the left and right lateral muscles and the interaction between body and fins and the water. Grillner and Kashin (1976) found in the eel that the *WoMO* down the body was 1.5 times faster than the *WoC*. From their graph showing the timing of the mechanical wave and the EMG, the velocities of both waves can be estimated. The speed of the *WoC* was 0.44 $m\,s^{-1}$ and the *WoMO* ran at 0.67 $m\,s^{-1}$ down the body (the eel was swimming at 0.41 $m\,s^{-1}$, body length was not given). The swimming speed of the eel of Table 8.1 and Fig. 8.5 was 0.28 $m\,s^{-1}$, the speed of the wave of curvature 0.40 $m\,s^{-1}$ and the velocity of the wave of bending was about 3.3 LT^{-1} or 1.7 $m\,s^{-1}$. So there was a greater than fourfold speed difference between the *WoB* and *WoC*. This would indicate that the *WoB* is faster than the *WoMO* in the eel. However, a comparison is only valid if steady swimming at uniform speed is compared in both cases. Comparison of the U/V values yields 0.95 for Grillner and Kashin's eel against a value of 0.69 shown in Table 8.1. Even 0.69 is a high U/V value for a swimming eel. A value of U/V of 0.95 indicates that the animal was probably not swimming at uniform speed but was slowing down, freewheeling through the water, while the film was made.

Signals from two EMG electrodes were recorded in a carp by Kashin *et al.* (1979) during slow swimming and a startle response. During slow swimming at a range of speeds the time lag between the start of the EMGs in the

Table 8.2 Kinematic and electromyographic data wave propagation along the bodies of swimming fish

Species	*Speed* U (LT^{-1})	WoC V (LT^{-1})	U/V –	WoMO V_m (LT^{-1})	*Phase-lag* at tail (T)	V_m/V –	*Source**
Lamprey		0.72		0.92	0.8	1.3	1
Trout		0.82		3.3/0.12	0.58		1
Eel	0.58	0.61	0.95	0.94	0.4	1.5	2
Tench				0.87			3
Carp	0.56	1.24	0.45	1.80		1.4	4
Saithe	0.86	0.98	0.87	1.55	0.77	1.6	5
Mackerel	0.83	1.02	0.81	1.55	0.74	1.5	5

*Sources: 1, Williams *et al.* (1989); 2, Grillner and Kashin (1976); 3, Blight (1976); 4, Van Leeuwen *et al.* (1990); 5, Wardle and Videler (1993).

front and rear remained more or less constant, independent of speed. The startle response result was interesting because the EMG activity was instantaneous at the two positions measured.

Williams *et al.* (1989) used EMG and film techniques to compare the wave characteristics of lamprey, *Lampetra fluviatilis*, and trout. The duration of the EMG bursts decreased towards the rear in both fish. The same happens but more extremely in carp (van Leeuwen *et al.*, 1990) and in saithe and mackerel (Wardle and Videler, 1993). The end of the EMG in these last three species occurs almost simultaneously all along one side of the body. A comparison of the existing quantitative EMG data (Table 8.2) is rather difficult owing to a lack of kinematic data. The *WoMO* is between 1.3 and 1.6 times faster than the *WoC*. The velocity of the *WoMO*, V_m, is constant along the body in carp, measured with eight approximately equally spaced electrodes on one side by van Leeuwen *et al.* and in the lamprey and eel, where four electrodes were used. There is one exception to this rule. The speed of both onset and end of muscle activity in trout was found to decrease substantially towards the rear by Williams *et al.*, using four electrodes.

The maximum curvature reaches the tail always later than the (extrapolated) *WoMO*. A phase lag in the caudal peduncle of 0.5 *T*, for example, would mean that the muscles on the opposite, convex side would start to be active when the maximum curvature on the concave side is reached.

In a study by Wardle and Videler (1993), 20 mackerel and 8 saithe, two pelagic fish species, were trained to swim between feeding lights in static seawater at 12 °C. The animals were fitted with two EMG electrodes at a distance of 0.40 *L* and 0.65 *L* on the left side. Analysis of steady swimming bouts at a range of speeds showed a surprisingly fixed relation between

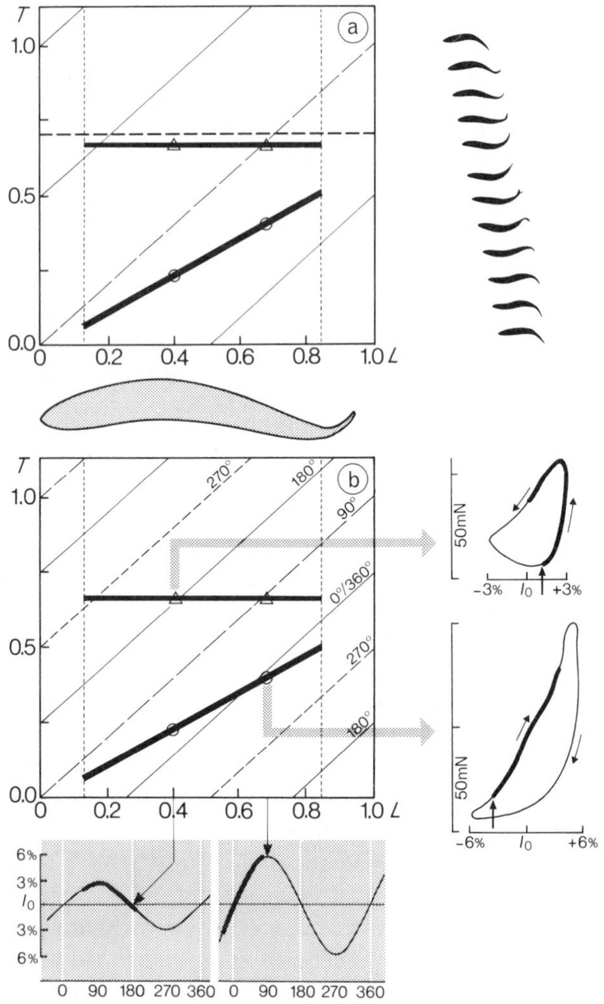

Fig. 8.6 (a) The onset and end of the EMG activity of mackerel and saithe related to the phase of maximum curvature on the right and left side of the body. The timing of the curvatures is essentially the same as in Fig. 8.5(a). The tail beat period *T* is used as the unit of time. Broken and solid oblique lines are the maximum curvature to the right and left respectively. Circles show times of EMG onset at about 0.40 *L* and 0.65 *L* and triangles indicate end of EMG activity. The black bars show possible onset and end extrapolated along the whole lateral muscle. The images to the right show swimming positions corresponding to the vertical time scale with leftmost tail positions at 0.0 *T* and 1.0 *T*. The two vertical dotted lines indicate the position of the front and rear ends of the lateral muscle. The silhouette of the mackerel in dorsal view below the horizontal axis is from an image at 0.75 *T*. The horizontal broken line corresponds with the instant of maximum power of the anterior muscles and maxi-

duration of EMG activity, velocity of the *WoMO* and the tail beat cycle. In both species, the duration of EMG activity at both electrodes remained a constant portion of the period of the *WoC* at all frequencies between 1.8 and 13 Hz. In mackerel and saithe respectively the onset of EMG activity at the front was at 0.74 *T* and 0.77 *T* before the leftmost tail position. The speed of the *WoMO* was 1.55 LT^{-1}. Fig. 8.6 gives a generalized picture for the mackerel (it would be about the same for saithe) of the timing of the EMG along the body in relation to the *WoC* and strain cycles of the muscles involved. The question remains of where in this picture the wave of maximum force production or bending moment (*WoB*) should be drawn.

Chapter 7 (Section 7.3) showed that force production in cyclically active muscles strongly depends on the timing of activation in relation to the phase in the strain cycle. The next section tries to relate the movements of the body, onset of muscle activity and optimal work loops of muscle fibres in swimming fish.

8.4 OPTIMUM WORK LOOPS OF MUSCLE FIBRES ALONG THE BODY

The basic principle of Fig. 8.6(a) is the same as used for the diagrams of Fig. 8.5. The horizontal distance between the two vertical axes represents the body length of the fish from head (left) to tail (right). The time is passing in an upward direction and the unit of time is the period *T* of the *WoC*. *T* is chosen to be zero when the tail position is leftmost. Maximum curvature to the left is indicated by the solid oblique lines, and maximum curvature to the right is represented by the parallel broken lines. The relative velocities of the *WoC* are typically close to 1 LT^{-1} for saithe and mackerel (Videler and

---

mum force in the posterior fibres. (b) The same diagram as in Fig. 8.6(a) but instead of indicating the waves of curvature, the diagonal lines represent phases in the strain cycle of the muscle fibres along the body. Line 0/360°, muscle fibres at resting length l_0, while lengthening; line 90°, fibres at maximum length; line 180°, fibres at l_0 while shortening; line 270°, muscle fibres pass through the minimum length. Strain cycles for the fibres at the electrode positions are drawn below the diagram. The strain of the anterior fibres is $\pm$ 3% l_0 and that of the posterior fibres $\pm$ 6% l_0. The corresponding work loops for these two positions are based on force measurements on isolated fibres from these positions in saithe by Altringham *et al.* (1993). (The preparations differed in cross-sectional area and length, so comparison of the absolute values of the magnitude of the forces and the surface areas is meaningless.) The thick parts of the lines on the work loops and of the lines representing the strain cycles indicate the duration of the EMG activity. Small arrows represent the instants of stimulation and long arrows indicate the direction of the work loops.

Hess,1984). The positions of the EMG electrodes, at about 0.40 *L* and 0.65 *L*, are indicated by circles and triangles. The circles show times of onset and the triangles indicate the end of the EMG activity. The thick bars show the expected timing of onset and end of the EMG, extrapolated along the whole lateral muscle. The vertical dotted lines indicate the span of the lateral muscle from just behind the head to the end of the caudal peduncle. The image below the diagram is the silhouette of a mackerel at 0.75 *T*. Mackerel silhouettes at right show swimming positions corresponding to the vertical time scale with leftmost tail tip positions at times *T*. Fig. 8.6(a) shows that the curvature to the left coincides with the end of the EMG signal in the front part of the body. Near the caudal peduncle muscles are active when the body is curving to the opposite side. We will take a closer look at what is actually happening at the muscle fibre level in Fig. 8.6(b). The diagram represents the same average swimming bout as Fig. 8.6(a), but the oblique lines indicate phases in the length-change cycles of the muscle fibres instead of maximum curvatures. Muscle fibres in the myotomes are subjected to sinusoidal length changes around the resting length l_0 during steady swimming. Examples, based on Altringham and Johnston (1989, 1990b) and Altringham *et al.* (1993), for the two electrode positions on the body are given below the phase diagram. During steady swimming, the muscle fibres near the front undergo length changes of $\pm$ 3% l_0. The EMG starts at about 30° and lasts until 180°. The muscle will stretch until it reaches l_0 + 3% at 90°, and is half-way through its shortening phase when the EMG ends at 180°. Experiments with fibres isolated from that region of the body of saithe gave maximum power output at all cycle frequencies when the first stimulus was given 30–40° after the muscle lengthened through l_0. Furthermore, the duration of the stimulus had to occupy about 160°, and the strain amplitude at higher frequencies had to be $\pm$ 3% l_0, to obtain maximum power output. These conditions are rather accurately met by the muscles near the front in Fig. 8.6. This implies that the work loop of the muscle fibres at that position must be close to the expected maximum. A work loop based on experiments on isolated muscle fibres from the same position of saithe tested under the optimal conditions described (strain $\pm$ 3% l_0, one stimulus at 30°; from Altringham *et al.*, 1993) is drawn in Fig. 8.6(b). It generates positive work during shortening. The maximum force occurs at about 100° of the cycle, and maximum power (force times velocity) is generated at about 180°.

The rear muscle fibres operate in a completely different way. Strain amplitudes are $\pm$ 6% l_0 owing to the higher amplitude of the lateral movement. EMG activity starts at 350°, 10° before the fibres reach their resting length while stretching. The duration of the EMG burst covers only 100° of the cycle and stops just before the maximum fibre length is reached at 90°. Experiments with isolated fibres demonstrated that under these conditions the highest forces were generated. A stimulus starting earlier or later in the cycle yielded lower forces. For example, stimulating at 40° generated only about half the force obtained with the same stimulus at 350°, for the same strain amplitude. The work loop from the caudal fibres is negative: the muscles resist stretching and stiffen the rear part of the body. Saithe muscle fibres taken from the caudal region at 0.65 L were tested *in vitro* by Altringham *et al.* (1993). The example work loop in Fig. 8.6(b) is reconstructed from their data. The measurements were made under strain conditions of $\pm$ 6% l_0, giving one stimulus at 330°. Maximum force occurs at 90°, i.e. at about 0.7 T, the same instant when the rostral muscle generates maximum power. This instant is only slightly earlier than the instantaneous maximum bending moment at 0.8 T (Fig. 8.5) predicted by Hess and Videler (1984) for saithe. It is indicated as a dashed horizontal line in Fig. 8.6(a).

The conclusion that muscles from front to rear operate in different ways to power the swimming movement in interaction with the water is now founded on theoretical and experimental evidence. This encourages speculations about how the bulk of muscles in the front part of the fish delivers forces to the tail blade where the main propulsive interaction with the water occurs.

8.5 FORCE TRANSFER FROM HEAD TO TAIL

The muscle function gradually changes from front to rear in swimming pelagic fish such as mackerel and saithe. The phase differences between caudally travelling waves of curvature and bending moment produce the differential use of fibres in the myotomes. Rostrally, muscle fibres generate power while shortening; caudally, active muscles resist stretching, by forces from cross-bridge formation enhanced by passive elasticity, generating large forces to withstand forces from the water pushing the tail blade. Between these extremes there is a transition zone where the function of the lateral muscles changes from power generators to force generators. Half-way down the body, muscles are active just before and after the instant when they reach their maximum length. They will do some negative and some positive work. The duration of the activity of the rostral myotomes is four times as long as that of the muscles in the caudal peduncle. The long contraction increases the amount of positive work and generates maximum force late in the cycle. The maximum tail beat frequency strongly depends on the contractile properties of these muscles. Caudal myotomes are only briefly active but at a crucial moment, namely when the maximum force is produced while the tail blade crosses the swimming track. At that instant, the power generated by the rostral muscles is maximal and is transferred efficiently along the muscles with increasing stiffness to the tail. This scenario

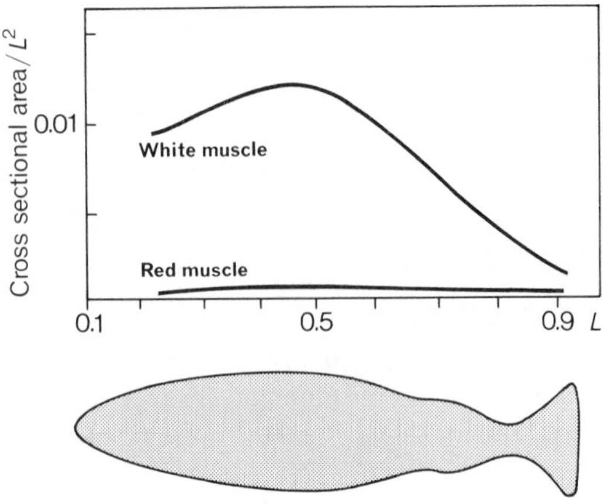

Fig. 8.7 Distribution of cross-sectional areas of red and white muscle of saithe along the body. Redrawn from He (1986).

requires hardly any other structure to explain the force transfer from front to rear.

However, one important aspect has not yet received proper attention. The total force generated by muscles is proportional to the surface area of the cross-section perpendicular to the fibre direction, and the total power is proportional to the muscle volume. Both cross-sectional areas and the volume occupied by the muscles decrease from head to tail in most pelagic fish. In saithe, for example, Fig. 8.7 shows this to be very evident for the white muscle and suggests that the red lateral muscles do not contribute to that effect. The white muscles in the first half of the body of saithe are able to deliver the largest bending power, whereas Fig. 8.4 indicates that the largest mean bending power per unit length is generated in the posterior part of the body. Blickhan (1992) came across the same discrepancy in trout. Maximum bending moments based on cross-sectional areas from local muscles are found in the anterior half of the body; actual bending moments generated by a live fish reached the highest values at the rear half of the body (Fig. 8.8) where they are more than twice as high as the bending moments that could be delivered by the local muscles. Part of this can be explained by the fact that muscles in the tail region operate during the stretching phase of the strain cycle, with forces that are up to 1.75 times higher than when activated in the same phase as the anterior muscles (Altringham *et al.* 1993). But even so, part of the large bending moments in the caudal region remains as yet unexplained.

The other structures capable of transferring forces from head to tail are the vertebral column and the skin.

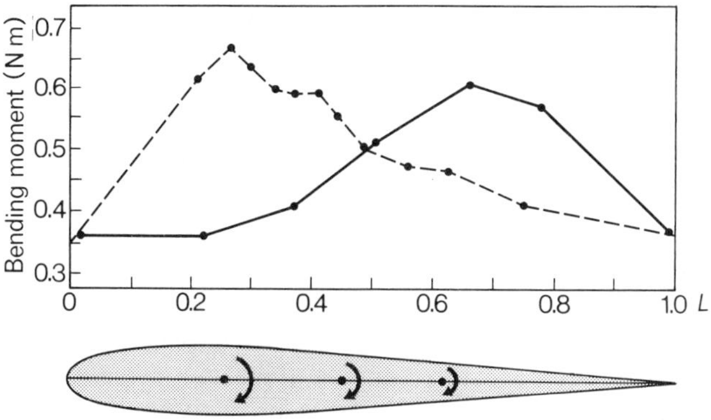

Fig. 8.8 Theoretical and actual bending moments along the body of a trout. Broken curve: maximum bending moments based on cross-sectional areas of the local muscles. Solid curve: actual bending moments measured by forcing a fish to push against a series of force transducers. Redrawn from Blickhan (1992).

8.6 BULGING MUSCLES AS AN ALTERNATIVE

During shortening, the thick and thin filaments slide along each other decreasing the distances between the Z-discs. The whole muscle shortens and thickens because the volume remains constant. This simple consideration implies that a muscle is able to generate force in more than one direction. The conventional conception is that muscles produce force in the fibre direction, either while shortening or by resistance to shortening against an external force. Muscular force transfer, however, could happen equally well crosswise, as the result of bulging, as well as lengthwise through shortening. It is not easy to see how bulging forces could be transduced.

The main bulk of muscles of typical pelagic fish, such as mackerel and saithe, is found in the first half of the body; the caudal peduncle contains only a small fraction of the total muscle mass. Measurements of the cross-sectional area occupied by red and white muscle of saithe are drawn in Fig. 8.7. The amount of red muscle is fairly constant from head to tail, in contrast to the volume of white fibres. The parallel red fibres probably power the slow swimming strokes by direct transfer of the force along the muscle. The main bulk of white fibres of the front part shortens while active and bulges. The lesser portion in the rear stretches and gets thinner and longer. The total volume of bulging muscles is much larger than the thinning lot in the caudal peduncle. The net effect is that the total span of the white muscle between head and tail shortens and that the skin in the thickest front half is inflated. (This argument does not apply to fish without a narrow caudal peduncle, such as eel.)

Owing to the oblique fibre direction of the collagenous fibres in the skin (Section 4.3), passive bending (not generated by the lateral muscles) will cause little or no stretch. Strain measurements during swimming in trout (Blickhan, 1992; Müller, 1992), eel (Hebrank, 1980) and lemon shark, *Negaprion brevirostris* (Wainwright *et al.*, 1978) show, however, that the skin is actually stretched when the body is bent by its lateral muscles. Preliminary results by Müller and Blickhan reveal that the strain in the fibre direction increases linearly with swimming speed, over a range of speeds where the body curvature remains the same. Evidently the bulging muscles stretch the skin and transfer stress forces through the skin of some species, but it is difficult to estimate to what extent. It is probably negligible in tunnid fishes with weak skins and strong tendons in the caudal peduncle, and more substantial in species without tendinous muscles and strong skins.

Stress–strain curves of pieces of fish skin consistently show by far the highest stress values per unit strain when tested in the fibre direction. The largest extensibility is found when the skin is tried in longitudinal direction. Intermediate stress and strain values occur by pulling dorsoventrally.

The collagenous fibres in the skin are wrapped around the fish as geodesic lines over the surface along the shortest distance between the dorsal and ventral edges of the median septum. Each fibre connects a frontal point on one edge with a more rearward one on the other edge. Forces along the fibre direction caused by the bulging frontal myotomes pull on the median septum and will try to bend the median septum, including the vertebral column, laterally. The rearward part of a typical pelagic fish is easier to bend than the thick frontal bit behind the head. The vertebral spines point obliquely in caudal direction and reinforce the median septum approximately in the pulling direction of the skin fibres. The angle between the vertebral column and the spines decreases to the rear in accordance with decreasing angles between the skin fibres and the longitudinal body axis. The fact that the dorsal and ventral spines are usually not long enough to reach the edge of the septum and are not directly connected to the skin argues against this hypothesis.

There is also the possibility that forces are transferred in longitudinal direction via the skin to the tail, despite the fact that uniaxial tests show that the skin is highly extensible in that direction. The skin could behave as a fabric, easily demonstrated with a handkerchief. It cannot be stretched much by pulling parallel to either of the fibre directions, but it can be stretched quite a lot by pulling along a diagonal, unless someone else pulls along the other diagonal. Bulging stretches the skin in dorsoventral direction along the vertical diagonal of the fibre directions. The skin acts like a fabric and will tend to shorten in the direction of the longitudinal diagonal (Videler,

1975). Biaxial experiments with lemon shark skin, testing the longitudinal stress and strain in the skin under constant dorsoventral stress, showed a substantial increase of the stiffness in longitudinal direction (Wainwright *et al.*, 1978). However, similar experiments with skin of eel, skipjack tuna and Norfolk spot failed to show this effect (Hebrank, 1980; Hebrank and Hebrank, 1986). Actual measurements of the forces in the skin of a swimming fish could probably show if the skin is used, at least by some species, as an external tendon transmitting forces from the main bulk of muscle to the tail. These experiments await the invention of suitable force transducers.

Anyone interested in this problem, central to our understanding of fish locomotion, is advised to read Wainwright's (1983) 'To bend a fish'.

8.7 SUMMARY AND CONCLUSIONS

Kinematic studies establish the temporal relations between cyclic events on the body of a swimming fish. Saithe and eel are used as examples. The velocity of the wave of curvature is about 1 LT^{-1} for saithe and 0.7 LT^{-1} for the eel. The amplitude of the lateral movements increases linearly from head to tail in the eel and follows a more complex power function on saithe.

Hydrodynamic theory allows estimates of reactive forces from the water on the fish, given the movements of the body and estimates of the mass of water affected by these movements as a function of the position on the body. The kinematic data also allow calculation of the internal forces and bending moments as functions of time and position on the body, given the division of body mass between head and tail. The maximum bending moment occurs instantaneously along the body of saithe at about 0.8 T if the leftmost tail tip position is taken as $T = 0$. The maximum bending moment on the eel runs at 3.3 LT^{-1}, 4.7 times faster than the wave of curvature.

The relative timing of the onset of electrical activity, measured from EMGs at several positions along the body of swimming fish, varies between species. The duration of the burst of activity generally decreases towards the rear. The speed of the wave of EMG-onset was at 1.55 LT^{-1} faster than the wave of curvature in mackerel and saithe.

Lateral muscle fibres of steadily undulating fish lengthen and shorten cyclically. The start of the activation and its duration are superimposed on the cyclic changes in length, and occur at different phases depending on the position on the body. In mackerel and saithe, the amplitude of the strain cycles is $\pm$ 3% of the resting length of the fibres in the anterior part of the body and increases to $\pm$ 6% at the caudal peduncle. The anterior muscles are activated while being stretched just before they reach the longest length and the activation lasts well into the shortening phase. *In vitro* experiments reveal that this timing provides a positive work loop, maximizing the generation of power. The posterior fibres are briefly active during the stretch phase of the cycles. This eccentric activity provides negative work loops and maximal forces.

The instants of maximum power generation in the front part, maximum forces in the rear part of mackerel and saithe coincide with the moment when the tail crosses the swimming track and delivers maximum propulsive force. The stiffened muscles in the caudal region are considered able to transfer most of the power generated by the main bulk of muscles in the front. The rest might be transferred through the skin and/or the vertebral column.

Shortening muscles will bulge because the volume has to remain constant. The bulging effect of the main bulk of frontal muscles could transmit forces through the tight helically wound collagenous fibres of the skin. Two hypothetical pathways are discussed but the experimental evidence is too scarce to prove either of these.

Chapter nine

The costs of swimming

9.1 INTRODUCTION

The costs of swimming concern the rates of energy expenditure needed to generate movements and forces for interacting with the water. Fish, like other living systems, use oxygen to combust stored energy-rich substrates (proteins, fats and carbohydrates) obtained from food. The energy released is needed for maintenance, growth, reproduction and activities including locomotion. As far as swimming is concerned, the oxidation process produces the energy required to make high-energy compounds such as ATP, which is the fuel used by muscles to generate force. This process also produces carbon dioxide, water and heat. In principle the rate of fuel depletion, the amount of oxygen used, the rate of carbon dioxide production or even the amount of heat produced could be exploited to estimate the energy turnover during swimming. The capacity of water to absorb heat and carbon dioxide is very high and the amounts of heat produced are relatively low. Despite these problems, Addink *et al.* (1989) measured heat production of a goldfish directly in a confined space, but were not able to measure increments during swimming activities. Under the same circumstances, fuel depletion in the muscles of an intact fish has been measured directly by the same research group using the method of nuclear magnetic resonance spectroscopy (NMR). Oxidation rates of a variety of substances in the blood of a cannulated swimming trout were measured by van den Thillart (1986).

Of the various options, the rate of oxygen consumption has been most commonly measured to estimate the costs of swimming. A variety of closed flume tanks, where fish are forced to swim against a flow of water at various speeds, has been designed for this purpose. The rate of depletion of oxygen is measured. Beamish (1978) gives an overview of these swimming chambers and experimental procedures.

Conversion of oxygen consumed to energy used is not a straightforward procedure because various substrates have different energy contents and use

Table 9.1 Energy equivalents used in fish bioenergetics (based on Brett and Groves, 1979)

Substrate	*Energy content*	*Oxygen used*	*Oxycaloric values (1 ml O_2 = 1.43 mg)*		*RQ*
	(J mg^{-1})	*(ml mg^{-1})*	*(J ml^{-1} O_2)*	*(J mg^{-1} O_2)*	–
Carbohydrate	17.2	0.81	21.1	14.7	1.00
Fat	36.3	1.85	19.6	13.7	0.70
Protein	20.1	1.05	19.2	13.4	0.90
Values used	–	–	20.1	14.1	–

different amounts of oxygen per unit weight to produce that energy. The ratio of CO_2 produced over O_2 uptake, the respiratory quotient (RQ), differs also between carbohydrate, fat and protein. Table 9.1 shows how much oxygen these substrates use to produce their specific amounts of energy as heat when burnt. It demonstrates that a prediction of the energy produced, based on oxygen consumption, requires knowledge of the substance oxidized. Extreme RQ values can give an indication of the substrate depleted, because pure carbohydrates yield 1 and pure fat combustion 0.7. Usually a mixture of substrates is exploited. An RQ of 0.96 has been measured for swimming trout (van den Thillart, 1986) and is generally accepted for fish swimming steadily under normal aerobic circumstances. The oxycaloric equivalent based on this RQ is 20.1 J per ml O_2 (or 14.1 J per mg O_2), and this figure will be used in all estimates presented here. The use of oxygen consumption to obtain an estimate of the energy used becomes extremely complicated if bouts of anaerobic swimming have to be taken into account.

Three metabolic rate levels can be recognized in fish. The basal metabolism is the minimum rate of energy expenditure needed to keep the fish alive. The standard metabolic rate includes also the extra energy needed to bring the animal to an increased activity level but does not comprise the energy needed to swim at a particular speed. The total energy used during swimming is commonly referred to as the active metabolic rate. The levels of these separate metabolic rates are species, size, temperature and velocity dependent. Substantial differences can be expected between, for example, a passive angler fish, *Lophius piscatorius*, and a fast active tuna. Condition and training also affect the results of measurements. The variation between the effects of size on standard levels is so large between species that no general relationship is valid. Standard levels vary about tenfold between 0.09 W kg^{-1} and 0.9 W kg^{-1} for fish between 10 and 100 g. The temperature effect on standard levels is probably best described with a $Q_{10\,°C}$ of somewhat higher than 2. There is considerable dispute and uncertainty about changes in standard levels with increasing speed (Stainsby *et al.*, 1980). There is probably no such thing as one standard level and I will therefore only use gross total levels of energy expenditure for comparisons between species.

In fish, active metabolic rates are directly proportional to mass. This is strange because one would expect larger animals to be relatively more efficient. Brett and Groves (1979) note that the relative amount of muscle in a large fish is higher than in a small fish (65% of the body mass in a 1000 g salmon and 35% in a 10 g salmon). The larger amount of active tissue probably prevents the expected decrease in metabolic rate. The highest levels measured in fish are about 4 W kg^{-1}. Fast streamlined fish can elevate metabolic rates with increasing speeds to up to 10 times the standard levels. This does not include increases obtained during anaerobic burst speeds, where levels of 100 times the standard rates can be achieved (Brett, 1972). The bodies of poikilotherm fishes are a lot cheaper to run than homeothermic mammals. For comparison, the highest active metabolic rate of a 2 kg sockeye salmon equals the basal metabolic rate of a rabbit of the same weight (Brett and Groves, 1979).

The drag of a steadily swimming fish is approximately proportional to the square of the speed. The energy needed to overcome that drag will therefore be nearly proportional to the cube of the speed. A satisfactory comparison between the costs of swimming is rather awkward to make owing to the large range of swimming speeds and hence metabolic rates even within each single animal. It requires a well-defined standard approach where size and speed are normalized and where temperatures are known. The next section of this chapter describes such an approach, which is subsequently used to compare the cost of locomotion of fish of a large range of sizes and with different swimming styles. The same method is employed to compare fish values with the swimming costs of other aquatic animals and with the energetic costs of flight and of terrestrial locomotion.

9.2 A FAIR COMPARISON BETWEEN COSTS OF SWIMMING

An unbiased comparison between costs of locomotion demands normalization of the speed. Maximum velocities and maximum sustained velocities have been used (e.g. Webb, 1975a). However, estimates of maxima are difficult to obtain and there is usually a fair amount of disagreement about the actual values (Wardle and Videler, 1980b). Tucker (1970) defined an optimum swimming velocity or maximum range speed (u_{opt}) where the amount of work per metre reaches a minimum. This optimum speed can be accurately calculated from measurements of energy expenditure over a range of speeds.

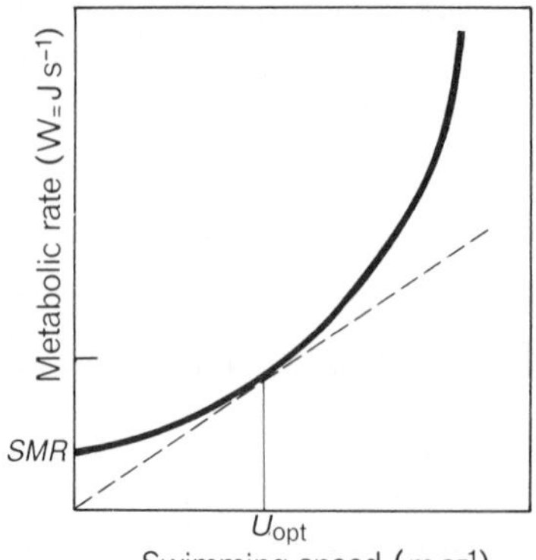

Fig. 9.1 Metabolic rate as a function of swimming speed. *SMR* is the standard metabolic rate at speed 0. The amount of work per unit distance covered is at a minimum at u_{opt}.

We expect the active metabolic rate (*AMR*) to increase with speed (u) as:

$$AMR = a + b\,u^x \qquad (x > 1) \qquad [W = J\,s^{-1}] \qquad (9.1)$$

where a and b are constants, and a is close to the standard metabolic rate (*SMR*). Fig. 9.1 provides an example of the expected shape of the relation between u in m s^{-1} and *AMR*. The ratio of *AMR* over u gives the amount of work per metre which reaches a minimum at u_{opt} where the tangent from the origin touches the curve. Values for the constants a and b and for the exponent x of Equation 9.1 can be found by iterative regression of published data until a least-squares solution is found which explains the largest proportion of the total variation. Whenever the energy expenditure at speed zero is not measured directly, the intercept, a, emerging from the regression of the swimming data, will be used as an estimate of the *SMR* instead. The amount of work per metre (*WPM*) can be found from Equation 9.1 by definition:

$$WPM = AMR\,u^{-1} = a\,u^{-1} + b\,u^{(x-1)} \qquad (J\,m^{-1}) \qquad (9.2)$$

Differentiation with respect to u gives:

$$WPM' = -a\,u^{-2} + (x - 1)\,b\,u^{(x-2)} \qquad (9.3)$$

which is zero when:

$$u_{opt} = \{a / [(x - 1)\,b]\}^{1/x} \qquad (m\,s^{-1}) \qquad (9.4)$$

Subsequent substitution in Equation 9.2 gives the minimal gross cost of locomotion. Tucker (1975) corrected the cost of locomotion (J m^{-1}) for size effects by dividing by body weight to obtain the dimensionless cost of transport. At the optimum speed, the minimum amount of energy required per unit weight and per unit distance ($\mathrm{J\,N^{-1}\,m^{-1}}$) is given by:

$$COT = AMR_{\mathrm{opt}}\,(M\,g\,u_{\mathrm{opt}})^{-1} \qquad (\mathrm{J\,N^{-1}\,m^{-1}}) \text{ or } (-) \qquad (9.5)$$

where AMR_{opt} is the active metabolic rate at u_{opt}, g the acceleration of gravity in $\mathrm{m\,s^{-2}}$ and M the body mass in kg. This dimensionless number provides a fair base for comparison. A 10 m basking shark, *Cethorhinus maximus*, however, covers 1 m in a fraction of one stride whereas small larvae would need many strides to cover that distance. The number of strides needed is approximately inversely related to the body length of the animal. Therefore, it is also useful to compare the costs of transport per unit weight over one body length ($COT \times L$) as well as the total energy needed for the animal to swim its body length ($COT \times L \times W$), which is equivalent to *WPM* times body length.

Two groups of data on fish swimming costs at the optimum speed are selected here (Table 9.2). A group of 14 species, where u_{opt} and *COT* could be calculated according to the method described, is considered to offer the most reliable data (Table 9.2 group A). A second group with nine species is based on papers where data allowed reasonable estimates but not precise determinations of u_{opt} and *COT*. In some cases, only routine swimming speeds were measured. Fish will most likely use optimum speeds during routine swimming and these speeds are therefore considered to represent u_{opt} (Table 9.2 group B). If optional, data near 15 °C are used. Most measurements were made in respirometer-type water tunnels or flume tanks (page 100).

9.3 FISH SWIMMING COSTS AT OPTIMUM SPEED

The optimum speed in $\mathrm{m\,s^{-1}}$ of group A of Table 9.2 is positively correlated with body mass (M in kg) according to:

$$u_{\mathrm{opt}} = 0.47\,M^{0.17} \quad (\mathrm{m\,s^{-1}})(r^2 = 0.86,\ N = 17) \qquad (9.6)$$

This is what one would anticipate, i.e. larger fish swim faster in $\mathrm{m\,s^{-1}}$ than smaller fish. The relation is, as expected, the other way round when the velocity is expressed in $L\,\mathrm{s^{-1}}$. The average relative optimum speed of group A is 2.3 $L\,\mathrm{s^{-1}}$, but the standard deviation (SD) is 1.3 $L\,\mathrm{s^{-1}}$. The variation in Table 9.2 group A extends from 5.8 $L\,\mathrm{s^{-1}}$ for the smallest Danube bleak, *Chalcalburnus chalcoides*, larvae to 0.8 $L\,\mathrm{s^{-1}}$ for cisco, *Coregonus artidii*. There is a negative correlation with an exponent of -0.14 between u_{opt} in $L\,\mathrm{s^{-1}}$ and body mass. This correlation is dominated by the figures for the larvae but

Table 9.2 Swimming costs at the optimum speed of fish

Species	*Mode**	T	M	L	u_{opt}		COT	*Source†*
		(°C)	(kg)	(m)	($m\ s^{-1}$)	($L\ s^{-1}$)	($J\ N^{-1}\ m^{-1}$)	
Group A‡								
Liza macrolepis	U	29	0.0083	0.105	0.21	2.0	0.67	1
Coregonus clupeaformis	U	12	0.364	0.340	0.46	1.3	0.18	2
Coregonus artidii	U	12	0.280	0.29	0.23	0.8	0.16	2
Oncorhynchus mykiss	U	15	0.264	0.292	0.28	1.0	0.29	3
Melanogrammus aeglefinus	U	10	0.156	0.248	0.25	1.0	0.22	4
Oncorhynchus nerka	U	15	0.0085	0.100	0.28	2.8	0.40	5
Oncorhynchus nerka	U	15	0.055	0.188	0.31	1.6	0.24	5
Morone saxatilis	U	15	0.212	0.254	0.43	1.7	0.32	6
Pomatomus saltatrix	U	15	0.225	0.254	0.51	2.0	0.36	6
Micropterus salmoides	U	15	0.150	0.225	0.42	1.9	0.21	7
Tilapia nilotica	U	25	0.080	0.21	0.41	2.0	0.21	8
Lepomis gibbosus	U	20	0.030	0.119	0.18	1.5	0.34	9
Cymatogaster aggregata	P	15	0.035	0.143	0.26	1.8	0.32	10
Rutilus rutilus	UL	20	5×10^{-6}	0.010	0.05	5.0	14.1	11
Rutilus rutilus	UL	20	3×10^{-4}	0.030	0.11	3.7	2.77	11
Chalcalburnus chalcoides	UL	20	1×10^{-5}	0.012	0.07	5.8	10.4	11
Chalcalburnus chalcoides	UL	20	5×10^{-4}	0.041	0.12	2.9	2.05	11

Group B‡								
Carassius auratus	U	15	0.100	0.18	0.12	0.7	0.20	12
Sphyrna tiburo	U	28	0.095	0.34	0.28	0.8	0.52	13
Sphyrna tiburo	U	28	4.65	0.95	0.48	0.5	0.17	13
Cymatogaster aggregata	P	15	0.03	0.12	0.14	1.2	0.39	14
Oxyjulis californica	P	15	0.07	0.17	0.20	1.2	0.35	14
Katsuwonus pelamis	U	23	0.6	0.35	0.56	1.6	0.33	15
Katsuwonus pelamis	U	23	3.8	0.6	0.72	1.2	0.30	15
Archosargus probatocephalus	P	28	0.35	0.3	0.99	3.3	0.15	16
Chilomycterus schoepfi	P	28	0.35	0.3	0.51	1.7	0.16	16
Cynoscion nebulosus	U	28	0.35	0.3	0.81	2.7	0.18	16
Sciaenops ocellata	U	28	0.35	0.3	0.90	3.0	0.24	16

*Column head abbreviations: Mode, swimming styles, listed below; *T*, water temperature; *M*, body mass; *L*, body length; u_{opt}, optimum velocity; *COT*, cost of transport at u_{opt}.

Swimming styles: U, lateral undulations of body and tail; P, pectoral fin swimmers; UL, undulatory movements of larvae.

†Sources: 1, Kutty (1969); 2, Bernatchez and Dodson (1985); 3, Webb (1971a, b); 4, Tytler (1969); 5, Brett (1964); 6, Freadman (1979); 7, Beamish (1970); 8, Farmer and Beamish (1969); 9, Brett and Sutherland (1965); 10, Webb (1975); 11, Kaufmann (1990); 12, Smit *et al.* (1971); 13, Parsons (1990); 14, Gordon *et al.* (1989); 15, Gooding *et al.* (1981); 16, Wakeman and Wohlschlag (1982).

‡For explanation of groups, see text.

nevertheless explains about 80% of the variation:

$$u_{opt} = 1.10\ M^{-0.14} \quad (L\,s^{-1})(r^2 = 0.80,\ N = 17) \qquad (9.7)$$

The variation is smaller in group B, where the average u_{opt} is 1.6 $L\,s^{-1}$ (SD = 0.9). Values of around 3 $L\,s^{-1}$ for 0.35 m long adult fish are probably too high. Equation 9.7 predicts u_{opt} values of 1.3 $L\,s^{-1}$ for these cases.

Weihs (1973b) predicts on theoretical grounds that a fish swims at its optimum speed when the energy needed for propulsion equals the rate of work ($J\,s^{-1}$) for standard metabolism. This theoretical optimum is at about 1 $L\,s^{-1}$, where the largest distance can be covered for a given amount of fuel.

Ultrasonic tag experiments with migrating sockeye salmon, *Oncorhynchus nerka*, in the ocean between Vancouver Island and mainland British Columbia showed that fish of an average length of 66.3 cm had an average speed of 66.7 $cm\,s^{-1}$, which is indeed about 1 $L\,s^{-1}$ (Quinn, 1988).

The dimensionless cost of transport at the optimum speed, compared in

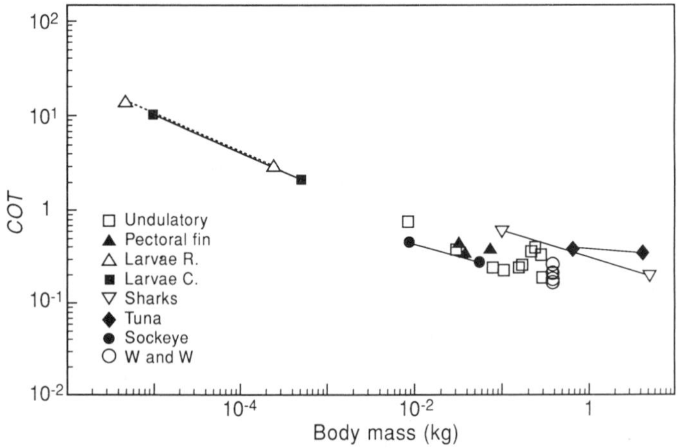

Fig. 9.2 Dimensionless *COT*, the energy needed to transport one Newton over one metre ($J\,N^{-1}\,m^{-1}$) during optimum-speed swimming, related to body mass. Points correspond to data of Table 9.2: heavy symbols are the data of group A, thin symbols belong to group B. Key: 'undulatory', species swimming with undulatory body motions (group A); 'pectoral fin', fish swimming with the pectoral fins (groups A and B); 'larvae R.', larval and juvenile roach, *Rutilus rutilus*; 'larvae C.', larval and juvenile Danube bleak, *Chalcalburnus chalcoides*; 'sharks', bonnethead sharks, *Sphyrna tiburo*, from group B; 'tuna', skipjack tuna, *Katsuwonus pelamis* from group B; 'sockeye', Brett's (1964) sockeye salmon, *Oncorhynchus nerka*; 'W and W', approximate data on four estuarine species from group B, published by Wakeman and Wohlschlag (1982). The points connected by lines are end points of series of measurements.

Fig. 9.2, shows a significant decrease with body mass:

$$COT = 0.11\ M^{-0.38} \qquad (-) \qquad (r^2 = 0.95,\ N = 17) \qquad (9.8)$$

This result is, of course, strongly influenced by Kaufmann's (1990) data on larval and juvenile cyprinids. The average regression coefficient for the larvae is -0.41. The pectoral fin swimmers appear to be just as economical as other fish. Sharks and the highly active tunas show the highest *COT* values relative to their body mass. Brett's (1964) classical data on sockeye salmon, found in most textbooks as prime examples of energy expenditure during fish locomotion, fit nicely with the more recent data. Bernatchez and Dodson (1985) found an interesting difference between two *Coregonus* species. The standard metabolic rate of lake whitefish, *C. clupeaformis*, was almost twice as high as that of cisco acclimated to the same temperature, but the net aerobic cost of swimming, obtained by subtracting standard metabolic rate from total oxygen consumption, was twice as high for cisco at any swimming speed. The total gross *COT* at u_{opt} of a 280 g cisco is lower than that of the 364 g lake whitefish, which is the opposite of the prediction from Equation 9.8. This indicates that differences in standard metabolic rates among species can have a considerable influence on the total cost of swimming.

Fig. 9.2 and Equations 9.7 and 9.8 also suggest that during ontogeny, both optimum speed in $L\,s^{-1}$ and the dimensionless cost of transport at that speed, gradually decrease.

Comparison of the cost to transport one unit weight over one body length offers a different picture in Fig. 9.3. There is no significant trend with body

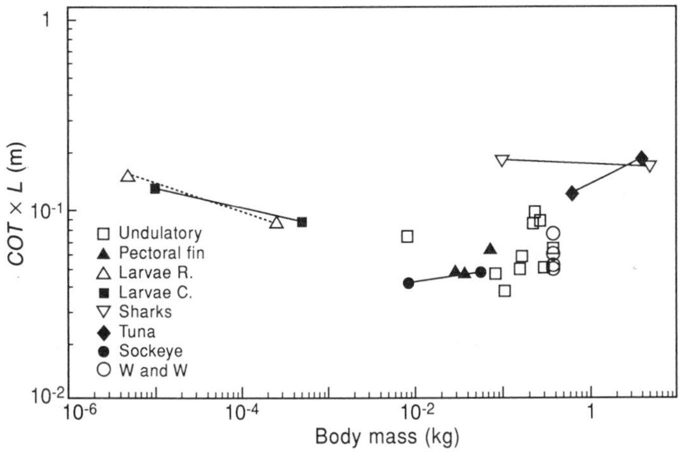

Fig. 9.3 Costs to transport one Newton over one body length, during optimum-speed swimming, related to body mass. Symbols as in Fig. 9.2.

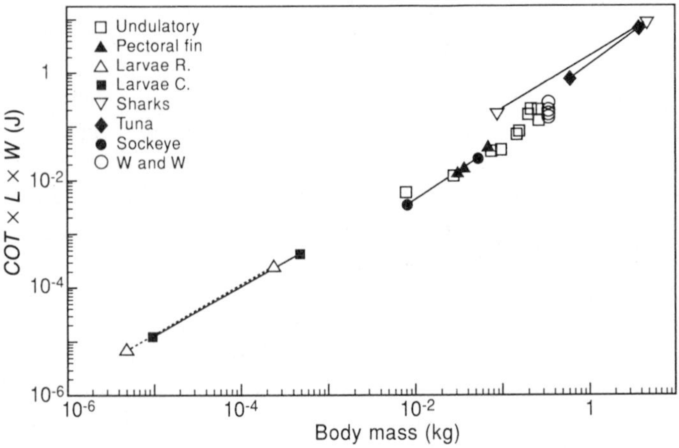

Fig. 9.4 Energy needed to transport the body weight of a fish at the optimum speed over its body length, related to body mass. Symbols as in Fig. 9.2.

mass. The values found for group A vary with an SD of 0.03 around an average value of 0.07 $\mathrm{J\,N^{-1}}$. Normalization of the distance covered brings cost of transport figures close together for a wide range of sizes. Each fish requires a similar amount of energy per unit weight to cover a distance equal to its body length. This result predicts the next.

The energy needed to transport the full weight of each fish over one body length at u_{opt}, or the absolute amount of joules needed, is almost directly related to body mass:

$$COT\ L\ W = 0.5\ M^{0.93} \quad \mathrm{(J)} \quad (r^2 = 0.99, N = 17) \qquad (9.9)$$

This equation and Fig 9.4, where the points of group B are presented as well, simply show that fish have to work harder to swim at the optimum speed when they grow bigger, despite the fact that the relative optimum speed (in $\mathrm{L\,s^{-1}}$) decreases with increasing size. The speed in absolute terms and the size of the animal increase; both factors increase the drag and hence the thrust required to overcome that drag.

9.4 FISH IN COMPARISON WITH OTHER SWIMMERS

A fair comparison between competitors requires knowledge of the handicaps involved. In Section 4.4 a sample of special adaptations showed that not all fish are adapted to be the most efficient swimmers at u_{opt}. Large parts of the variation in, for example, Fig. 9.3 are probably caused by differences in the importance of swimming at slow cruising speeds in the energy budgets of different species. Evolutionary pressures have not been focused on optimizing swimming efficiency at slow speeds or minimizing the cost of transport equally in all species. Larger variations can be expected if other aquatic animals are drawn into the competition. Some groups, e.g. cetaceans, pinnipeds, turtles and penguins, are secondarily adapted to live in water. Others, mink and humans for example, only swim occasionally. These last two are confined to swimming at the surface, which makes the comparison with submerged swimming animals rather unfair.

Swimming at the surface

Fish usually swim underwater and not near the surface. Air-breathing aquatic animals have to be at the surface at regular intervals. Whales and dolphins usually dive again after taking a breath of air because swimming at the surface causes extra drag by generating waves. A stiff, streamlined body just touching the interface between air and water experiences five times as much drag as the same body at a depth of more than three times its width. Fig. 9.5 shows this as the results of Hertel's (1966) experiments with a towed, spindle-shaped object. Fish experience drag penalties in the same order of magnitude when they are breaking the surface occasionally, for example during feeding or spawning activities. Webb *et al.* (1991) forced trout to swim near the surface where they indeed lost four-fifths (80%) of their swimming energy by generating waves. Waves are in fact masses of water lifted up against gravity. In shipbuilding practice, the relative amount

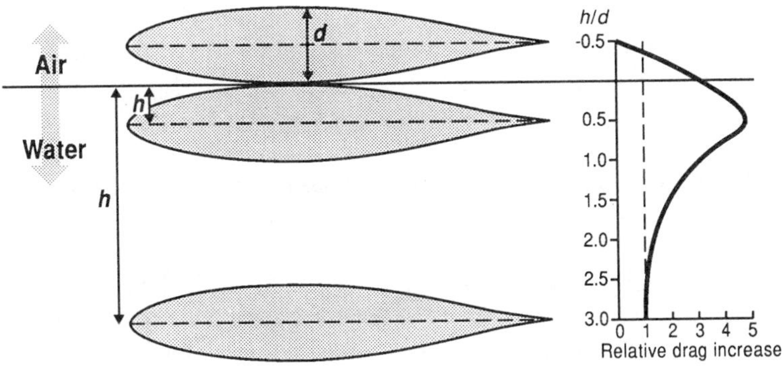

Fig. 9.5 Drag on a streamlined body (largest diameter, *d*) as a function of the submerged depth (*h*). Based on Hertel (1966).

of energy lost by wave generation is expressed by the dimensionless Froude number F:

$$F = U / \sqrt{gL} \qquad (-) \qquad (9.10)$$

where U is the speed of travel, g the acceleration due to gravity and L the length of the water-line. When F numbers are small, e.g. at low velocities and large length, only low waves are produced. Drag resulting from the generation of waves reaches maximum values at an F value of about 0.45. Above this value, the drag reduces because the speed becomes high enough for a boat to plane. For swimmers and boats that are not able to skim the water surface, $F = 0.45$ forms an unsurpassable limit. More propulsive energy will only produce higher waves, not higher velocities. The world record 100 m free style by A. Popov took slightly more than 48 s (at approximately 1 $L\,s^{-1}$). The F value during that event was about 0.45, which means that it will be virtually impossible for a person of the same length to swim substantially faster. I expect that the next world record will be swum by a longer person.

The energy used at the optimum speed of the widest available variety of swimmers, using different swimming styles, will now be compared with the performances of fish.

Comparing fish with submerged and surface swimmers

The data on COT at u_{opt} of submerged and surface swimmers were collected from the literature following the same selection criteria used for fish (Videler and Nolet, 1990). The data are compiled in Table 9.3 and visualized in the log–log plots of Figs 9.6, 9.7 and 9.8. The data of the surface swimmers are too scattered to provide significant trends. I will concentrate therefore mainly on the underwater swimmers for comparisons with the trends found in fish and use for this purpose specific data of surface swimmers only.

The u_{opt} in $m\,s^{-1}$ of the submerged swimmers is, as in fish, positively correlated with body mass:

$$u_{opt(sub)} = 0.47\, M^{0.29} \qquad (ms^{-1})(r^2 = 0.91,\ N = 17) \qquad (9.11)$$

The constant happens to be the same as in Equation 9.6, which means that any 1 kg animal, swimming underwater, is predicted to have a u_{opt} of about 0.5 $m\,s^{-1}$.

The average relative u_{opt} was 1.2 $L\,s^{-1}$ (SD 0.4) and 1.4 $L\,s^{-1}$ (SD 0.7) for submerged and surface swimmers respectively. Unlike fish, no significant relation between u_{opt} in $L\,s^{-1}$ and body mass could be found for the underwater swimmers, even though the shrimp data definitely seem to have such a relationship. That is, however, caused by the fact that they have a similar u_{opt} of about 0.45 $m\,s^{-1}$ and are of a range of different lengths from 2.7 to 6.9 cm.

Table 9.3 Cost of transport at u_{opt} of submerged and surface swimmers

Species	M *(kg)*	L *(m)*	u_{opt} *(m s⁻¹)*	*(L s⁻¹)*	COT *(J N⁻¹ m⁻¹)*	*Source**
Submerged						
Chelonia mydas	0.735	0.24	0.26	1.1	0.38	1
Chelonia mydas	1.15	0.29	0.49	1.7	0.31	2
Tursiops truncatus	145	2.25	4.00	1.8	0.14	3
Eschrichtius robustus	15000	11.50	2.25	0.2	0.04	4
Zalophus californianus	22.5	1.31	1.66	1.3	0.26	5
Eudyptula minor	1.2	0.40	0.70	1.8	1.29	6
Spheniscus demersus	3.17	0.65	0.86	1.3	0.78	7
Phoca vitulina	42.5	1.25	1.25	1.0	0.25	8
Phoca vitulina	33.0	1.30	1.61	1.2	0.37	8
Phoca vitulina	63.0	1.50	2.08	1.4	0.23	9
Palaemon adspersus	0.00012	0.027	0.045	1.7	8.21	10
Palaemon adspersus	0.00028	0.035	0.044	1.3	5.83	10
Palaemon adspersus	0.00052	0.043	0.046	1.1	5.31	10
Palaemon adspersus	0.00076	0.048	0.044	0.9	4.77	10
Palaemon adspersus	0.00143	0.059	0.046	0.8	4.10	10
Palaemon adspersus	0.00235	0.069	0.045	0.7	3.65	10
Loligo opalescens	0.041	0.20	0.37	1.9	1.27	11
Surface						
Amblyrhynchus cristatus	2.98	0.91	0.22	0.2	0.36	12
Eudyptula minor	1.20	0.40	0.72	1.8	1.62	6
Homo sapiens	76.3	1.79	0.28	0.2	1.05	13
Mustela vison	1.10	0.58	0.70	1.2	4.54	14
Aythya fuligula	0.613	0.27	0.41	1.5	1.71	15
Aythyla fuligula	0.613	0.27	0.53	2.0	1.82	16
Anas platyrhynchos	1.08	0.33	0.48	1.5	2.70	17
Anas superciliosa	1.10	0.33	0.72	2.2	2.55	6
Branta leucopsis	1.74	0.36	0.62	1.7	1.20	18

*Sources: 1, Prange (1976); 2, Butler *et al.* (1984); 3, Williams *et al.* (1991); 4, Sumich (1983); 5, Feldkamp (1987); 6, Baudinette and Gill (1985); 7, Nagy *et al.* (1984); 8, Davis *et al.* (1985); 9, Craig and Påsche (1980); 10, Ivlev (1963); 11, O'dor (1982); 12, Vleck *et al.* (1981), Gleeson (1979); 13, Nadel *et al.* (1974); 14, Williams (1983); 15, Woakes and Butler (1983); Woakes and Butler (1986); 17, Prange and Schmidt-Nielsen (1970); 18, Nolet *et al.* (1992).

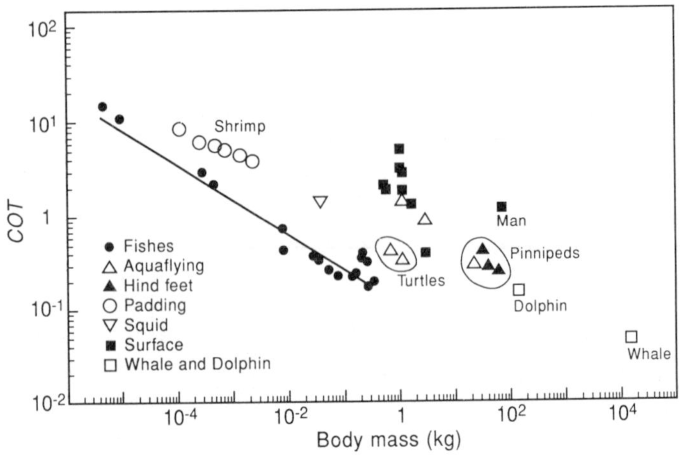

Fig. 9.6 Dimensionless *COT*, the energy needed to transport one Newton over one metre ($J N^{-1} m^{-1}$) during optimum-speed swimming, related to body mass of fish, submerged and surface swimmers. Points correspond to data of Table 9.3. Key: 'fishes', fish species of Table 9.2 group A, the regression line is found by log–log regression of the data described by Equation 9.7; 'aquaflying', animals swimming with the front legs (turtles, penguins and sea lion); 'hind feet', three data points of harbour seals, *Phoca vitulina*; 'shrimp', cost of paddling with swimmerets of *Palaemon adspersus* at different sizes measured by Ivlev (1963); 'squid', *COT* of jet propulsion of *Loligo opalescens*; 'surface', a variety of surface swimmers, the data are compiled in the lower part of Table 9.3; 'whale and dolphin', data points for the gray whale, *Eschrichtius robustus*, and the bottle-nosed dolphin, *Tursiops truncatus*.

The dimensionless *COT* plotted against mass in Fig. 9.6 compares the fish results with the cost to transport 1 N over 1 m of the swimmers varying from shrimps to the gray whale. The submerged swimmers show a clear decrease of *COT* with body mass:

$$COT_{sub} = 0.65\, M^{-0.27} \qquad (-)(r^2 = 0.95,\ N = 17) \qquad (9.12)$$

The decline is less steep than that of fish and the values are consistently higher. The whale, the bottle-nosed dolphin, the pinnipeds and the squid are approximately on the line through the *Palaemon* points. Two turtles are below that line and use almost as little energy as fish. Seals, swimming by hind feet oscillation, and the aquaflying sea lion have similar *COT* at u_{opt}. Conversely, a similar swimming mode like aquaflying yields highly different *COTs* between animals of different taxonomic groups. We already saw that pectoral fin swimmers among fish do not deviate in this respect from undulating fish. Turtles and the sea lion score relatively low using this technique but penguins are surprisingly high. The squid, *Loligo opalescens* and the heaviest sockeye salmon share approximately the same mass, u_{opt} and body length, but the squid, using jet propulsion, needs more than five times as much energy. The three reptiles in the data set show similar results despite the fact that the marine iguana, *Amblyrhynchus cristatus*, is swimming at the surface and the green turtles, *Chelonia midas*, submerged. The iguana is the cheapest surface swimmer: it uses half the energy of the submerged swimming penguin, *Spheniscus demersus*, swimming at a quarter of the speed of the bird of approximately the same mass. The mink, *Mustela vison*, is in terms of energy the worst surface swimmer. It uses per Newton and per metre almost twice as much as the duck, *Anas superciliosa*, at the same speed and with the same body mass. Man, at an optimum speed only one-seventh that of a seal of the same mass, uses just over four times as much energy. The *COT* of the little penguin, *Eudyptula minor*, swimming at the surface, is about 1.3 times as high as during submerged swimming.

The $COT \times L$ values are compared in Fig. 9.7. Fishes and turtles are by far the most economical, if we compare the costs to transport one Newton over one body length. All other submerged swimmers, no matter what their body mass, have higher costs. The differences are, however, remarkably small, the average value being 0.30 J N^{-1} (SD 0.13, N = 17) which is 4.3 times as high as that of the fish cluster. The surface swimmers, apart from mink and Man, do not yield substantially higher values. These are 0.6 J N^{-1} on average, which is more than eight times the figure found for the fish cluster. In contrast, mink needs 2.6 and Man slightly less than 2 J N^{-1} body weight to swim their body length. On the left-hand side of Fig. 9.7, the values found

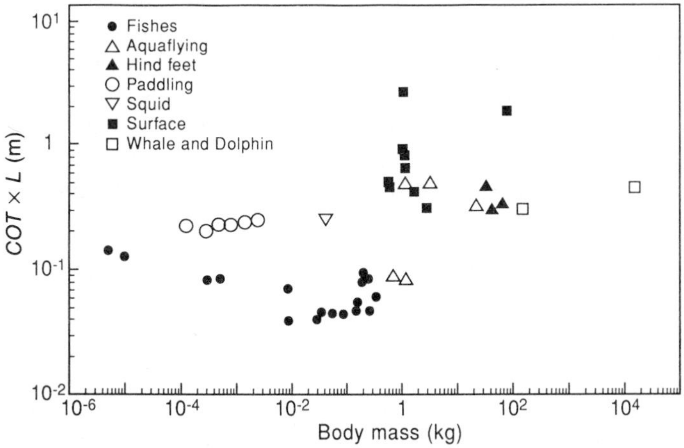

Fig. 9.7 Costs to transport one Newton over one body length, during optimum-speed swimming, related to body mass. Symbols as in Fig. 9.6.

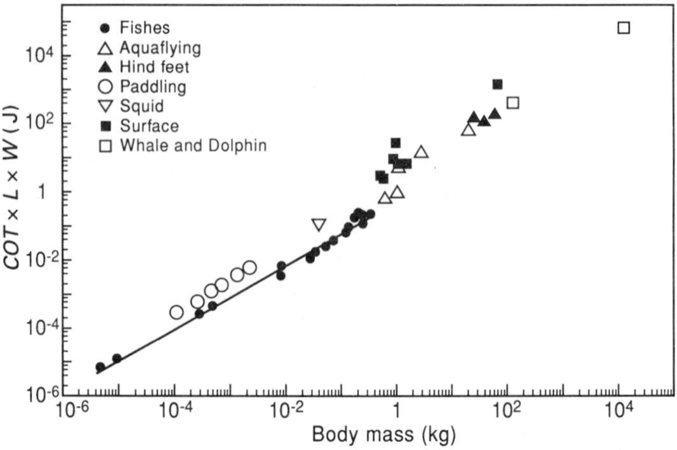

Fig. 9.8 Energy needed to transport the body weight of swimmers at the optimum speed over their body length, related to body mass. Symbols as in Fig. 9.6.

for the smallest animals seem to point at a common value of about 0.2 J N^{-1} for a body mass of approximately 1 mg.

The energy needed to transport the weight of each animal over its body length increases with body mass (Fig. 9.8). Fish and turtles are on a line, described by Equation 9.9, approximately parallel to but at one-fifth of the other submerged swimmers. The equation for the submerged swimmers is:

$$COT \times L \times W = 2.7\ M^{1.04} \qquad \text{(J)}(r^2 = 0.99,\ N = 17) \qquad (9.13)$$

Values for the surface swimmers other than mink and Man are in good accordance with this equation. Man needs 6.6 times the energy required by the largest seal to swim his body length and the mink four times as much as the duck of the same mass. The squid uses four times as much energy as the largest sockeye salmon which is about as long, of similar weight and swims at about the same u_{opt}.

The active metabolic rates (*AMR*), are found by multiplying *COT* by body weight and u_{opt}, in each case. Regression on body mass yields:

$$AMR = 0.5\ M^{0.80} \qquad \text{(W)} \quad (r^2 = 0.98,\ N = 17) \qquad (9.14)$$

for the fish and

$$AMR = 3.0\ M^{1.01} \qquad \text{(W)} \quad (r^2 = 0.99,\ N = 17) \qquad (9.15)$$

for the submerged swimmers. The energy consumption at u_{opt} roughly equals three times the figure for body mass for the submerged swimmers. Fish are cheaper in two ways: the increase in metabolic rate with body mass is less and the constant is a factor of 6 lower.

A comparison between swimming, flying and running

Most student textbooks on physiology use Schmidt-Nielsen's (1972) graph, showing that the cost of locomotion, the amount of energy required to move one unit of distance, is lower for swimming than for flying and running, despite the high viscosity and density of the medium water. Schmidt-Nielsen explains this seeming contradiction in terms of the relatively low speed and the streamlined, near neutrally buoyant bodies of swimming animals. His data on submerged swimming were largely based on Brett's (1964) studies of the sockeye salmon at 15 °C swimming at 75% of its maximum cruising speed. The comparison presented here is between cost of swimming at the optimum speed for each animal. The results for fish differ from those of most other submerged swimmers.

The cost of flight at uniform speed is high at low velocities, decreases with increasing speed to a species- and mass-specific minimum value, and increases from there approximately proportional to the cube of the velocity. The theoretical *AMR* curve for flying animals is therefore not J-shaped as that of swimmers shown in Fig. 9.1, but U-shaped (in practice measurements often deviate from this ideal curve). A U-shaped curve implies that there are two optimum velocities, one where the *AMR* reaches the minimum value and a higher speed where, comparable with the u_{opt} of swimmers, the amount of work per metre distance covered is at a minimum. This speed is called the maximum range speed (V_{mr}), and will be used here for the comparison. I searched the literature for data allowing the calculation of the *COT* at V_{mr} of flyers of zoological and technical origin, and compiled the results in Table 9.4 (for details of the approach see Videler, 1992). The *COT* values of insects, birds, bats and aircraft are displayed in Fig. 9.9 together with the regression lines for running animals, for fish and for the submerged swimmers. Log–log regression of the data points for the animal flyers yields:

$$COT_{\text{animal flight}} = 0.6M^{-0.25} \qquad (-)(r^2 = 0.81, N = 28) \qquad (9.16)$$

This is extremely close to the regression line of the submerged swimmers with a coefficient of −0.27 and a constant of 0.65 (Equation 9.12). The decline with mass of the *COT* of fish is steeper with a coefficient of −0.38 and the costs of fish are substantially lower over the mass range shown. Aircraft are at least six times as costly as the flying animals and the submerged swimmers.

A comparison with terrestrial locomotion is less straightforward, because the metabolic cost of running increases linearly with speed. This means that the amount of energy per metre distance covered is constant, and accordingly that there is by definition no optimum speed. The regression line is based on a comparison of the net cost of transport of running lizards, birds

Table 9.4 Dimensionless costs of transport at the maximum range of insects, birds, bats and aircraft (from Videler, 1992)

Species	*Mass (kg)*	V_{mr} *(m s^{-1})*	COT *(J N^{-1} m^{-1})*	*Source**
Drosophila (fruitfly)	7.20e−07	1.30	8.50	1
Simulium (blackfly)	2.30e−06	1.05	7.60	1
Aedes (mosquito)	2.50e−06	0.99	8.70	1
Tabanus (horse fly)	1.71e−04	2.20	4.20	1
Apis (honey bee)	8.00e−05	4.30	12.21	2
Apis (honey bee)	8.00e−05	4.30	13.99	3
Apis (honey bee)	8.40e−05	2.06	9.95	1
Apis (honey bee)	1.40e−04	4.30	7.99	3
Bombus (bumblebee)	5.00e−04	4.00	19.33	4
Schistocerca (locust)	2.00e−03	3.00	3.43	5
Calypte (hummingbird)	3.00e−03	13.61	1.72	6
Colibri thalassinus (green violetear)	5.50e−03	10.80	3.55	7
Colibri coruscans (sparkling violetear)	8.50e−03	8.00	3.14	7
Colibri coruscans (sparkling violetear)	8.50e−03	10.80	3.11	7
Melopsittacus (budgie)	3.50e−02	11.70	1.03	8
Sturnus (starling)	7.28e−02	16.80	0.75	9
Sturnus (starling)	7.50e−02	16.80	1.02	10
Phyllostomus (bat)	9.30e−02	8.00	1.23	11
Falco (kestrel)	2.00e−01	9.00	0.75	12
Corvus (fish crow)	2.75e−01	11.00	0.83	13
Larus (laughing gull)	3.22e−01	12.60	0.55	14
Columba (pigeon)	3.30e−01	12.00	0.65	15
Columba (pigeon)	3.84e−01	16.11	0.41	16
Columba (pigeon)	4.42e−01	10.00	0.61	17
Corvus (raven)	4.80e−01	11.00	0.64	18
Pteropus (fruit bat)	6.49e−01	8.60	0.72	19
Pteropus (fruit bat)	7.70e−01	8.00	0.74	19
Pteropus (bat)	7.80e−01	9.90	0.62	11
Cherokee	9.78e+02	55.56	0.51	20
Sikorsky S62 (helicopter)	3.59e+03	41.11	1.51	20
Grand Commander	3.86e+03	103.06	0.35	20
Jet fighter (F105F)	1.72e+04	319.44	0.65	21
Fokker 50	1.90e+04	145.00	0.28	22
Fokker 100	4.00e+04	165.00	0.30	22
DC9–10 (transport)	4.13e+04	244.17	0.27	20
DC8 (transport)	1.07e+05	268.06	0.24	21
Boeing 737–400	5.00e+04	220.00	0.25	23
Boeing 747	2.50e+05	250.00	0.19	23
Airbus 310	1.00e+05	235.00	0.21	23

*Sources: 1, Hocking (1953); 2, Nachtigall *et al.* (1989); 3, Wolf *et al.* (1989); 4, Ellington *et al.* (1990); 5, Weiss-Fogh (1952); 6, Lasiewski (1963); 7, Berger (1985); 8, Tucker (1968, 1972); 9, Torre-Bueno and Larochelle (1978); 10, Westerterp and Drent (1985); 11, Thomas (1975); 12, Masman and Klaassen (1987); 13, Bernstein *et al.* (1973); 14, Tucker (1969, 1972); 15, Rothe *et al.* (1987); 16, LeFebvre (1964); 17, Butler *et al.* (1977); 18, Hudson and Bernstein (1983); 19, Carpenter (1975); 20, Stanfield (1967); 21, Taylor (1968); 22, Fokker, aircraft brochures; 23, Data from performance manuals of aircraft used by KLM, Royal Dutch Airlines, kindly provided by Ir. W. Brouwer, Director Flight Technical Department (1991).

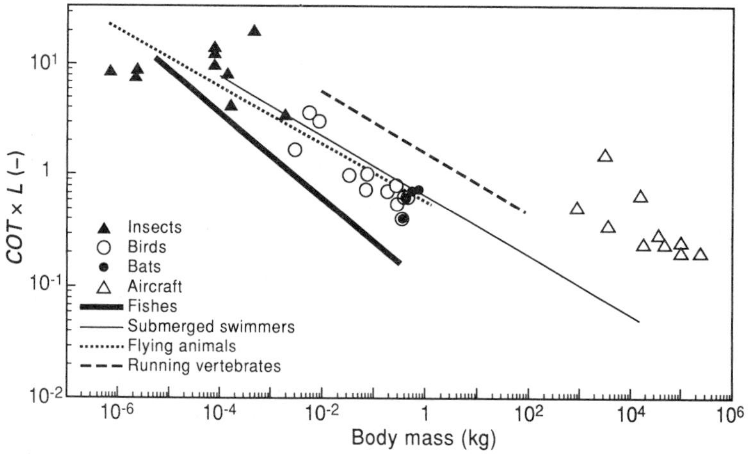

Fig. 9.9 A comparison between costs of transport at u_{opt} for swimmers, at the maximum range speed for flyers and at any speed for running vertebrates. The data for the flying animals and aircraft are listed in Table 9.4 and are based on Videler (1992). The running vertebrates regression line is based on data for lizards, birds and mammals from Fedak and Seeherman (1979). The regression line for fishes is the same as the line shown in Fig. 9.6. The regression line for the submerged swimmers is described by Equation 9.12. This line is not significantly different from a regression of the points for the flying animals. The two most expensive aircraft are a helicopter and a jetfighter respectively.

and mammals published by Fedak and Seeherman (1979), assuming the mass-specific standard metabolic rate to be proportional to 4.24 $M^{-0.25}$ (Kleiber, 1961). This is a rough approximation because *SMR* is velocity dependent (Peters, 1983). The slope of the line is not affected by this approximation. It is -0.28, which is almost equal to the slope of the line for submerged swimmers and to that of the flyers.

This improved analysis of energy costs of locomotion, comparing animals at physiologically equivalent speeds, still supports the main conclusion that fish have the cheapest mode of transport. Submerged swimmers and flying animals move about at similar minimum costs. They use more energy per Newton and per metre than fish but less than running vertebrates. Commercial aircraft are very expensive to run.

9.5 SUMMARY AND CONCLUSIONS

Fish usually burn a mixture of proteins and carbohydrates to release the energy for swimming. An RQ of 0.96 is commonly accepted for fish operating under aerobic conditions, where 1 ml of oxygen provides 20.1 J of energy. The combustion process could potentially be studied by measuring fuel depletion, oxygen consumption, carbon dioxide or heat production. The rate of oxygen consumption is predominantly used to estimate the energetics of swimming in closed flow channels or respirometers.

The cost of swimming is the sum of the standard metabolic rate and the energy required to produce thrust, to accelerate and to overcome the drag forces. It increases with speed following a J-shaped curve. The exact shape of this curve depends mainly on the species, size, temperature and the condition of the fish. To make fair comparisons possible, the optimum speed (u_{opt}), where the amount of energy used per unit distance covered is at a minimum, is used as a bench-mark. Series of measurements of oxygen consumption at a range of speeds provide the parameters needed to calculate u_{opt} and the energy used at that speed. The energy values were normalized by dividing the active metabolic rate at u_{opt} by the weight of the fish times u_{opt}, to reach a dimensionless number for the cost of transport (*COT* expressed in $J N^{-1} m^{-1}$). Hence, *COT* represents the cost to transport one unit of weight over one unit of distance. By multiplying *COT* by the length of the animal, the cost to transport 1 N over one body length could be compared. Multiplication of this last figure by the weight of the fish in N provided an estimate of the amount of energy needed to transport the weight of the animal over its length.

A literature survey yielded 14 species (17 single measurements) where u_{opt} and *COT* could be calculated and another group of nine species where fairly precise estimates of these parameters could be made.

The u_{opt} in $m s^{-1}$ is positively correlated with mass; u_{opt} decreases with mass if it is expressed in $L s^{-1}$ (the exponents are 0.17 and -0.14 respectively). The *COTs* are negatively correlated with body mass with an exponent of -0.38. There is no significant trend with body mass if the cost to transport 1 N over the body length of each fish is taken into consideration. Fish use on average 0.07 $J N^{-1}$ to swim their body length at u_{opt}. If the weight of the animals is taken into account as well, the amount of energy used to swim at u_{opt} increases in proportion to body mass (with an exponent of 0.93).

In a comparison with other swimmers, animals swimming at the surface have an extra handicap and should be treated separately. At the surface, considerable energy is turned into waves. Drag can be up to five times higher at the surface than deep down. The relative amount of energy lost this way is expressed by the Froude number.

The *COT* of submerged swimmers also decreases with body mass, but at a lower rate than that of fish, with an exponent of -0.27 instead of the -0.38 found for fish. The level is about five to six times as high as that of fish. Turtles are exceptional, with *COT* values close to the range found for fish. Surface swimmers are not as bad as expected. Man is the poorest swimmer, using extreme amounts of energy at a low u_{opt}. The average $COT \times L$ value for submerged swimmers is $0.30\ \mathrm{J\,N^{-1}}$, 4.3 times as high as that of fish. The total amount of work done by submerged swimmers increases linearly with body mass at a level about five times as high as that of fish.

A comparison between the *COT* of swimming and flying at optimum speeds and running at any speed shows similar figures for flying animals and submerged swimmers. Adult fish are considerably cheaper, and running is the most expensive form of animal transport, but it is more economical per unit weight and distance than flight by commercial aircraft.

Chapter ten

Ecological implications

10.1 INTRODUCTION

Swimming speed limits and endurance are directly related to food capture, escape from predators and reproduction. Therefore they are presumed to be subjected to strong selection pressures that enhance evolutionary fitness. This Darwinian fitness requires an individual not only to survive but also to produce fertile offspring. Against this background, all aspects of swimming performance are potentially crucial. The maximum speed a fish can sustain indefinitely, the endurance at higher speeds and the absolute maximum all-out burst speed are of high ecological importance and therefore important to investigate. It is also crucial to know how much time and energy must be allocated to swimming in order to achieve the highest fitness and how this affects the energy budget of a fish.

The first part of this chapter compiles the available data on swimming performance measured under experimental conditions and tries to find generally valid rules and estimates for the different speed limits and endurances. It ends with an example showing how the energetic costs of swimming are related to the limits of performance.

In the second part, the allocation of time and energy is the central theme. Examples show how the relative time spent swimming under natural conditions of an individual fish can be measured. Swimming is mainly needed for the acquisition of food and for activities serving reproduction. Swimming is therefore usually not an independent item of the time budget. Each of the items on the time budget must be translated into energetic terms (costs and benefits) to turn the time budget into an energy budget. A schematic energy budget illustrates the position of swimming costs on the expenses side of the balance sheet. Examples of some of the reciprocal relations, including the partitioning of energy, between swimming, food acquisition, diet, growth and reproduction will be given.

10.2 SPEED AND ENDURANCE

Maximum swimming records have attracted considerable attention. However, slower swimming speeds and the stamina at these speeds represent equally important survival values for a fish. Brett (1964) described the general form of the relation between endurance time and velocity in 0.18 m sockeye salmon tested at 10 °C. At low speeds these fish can swim continuously without showing any signs of fatigue. Limited endurance can be measured at speeds higher than the maximum sustained speed (U_{ms}) of about 3 $L\,s^{-1}$. For these prolonged speeds, the logarithm of the time to fatigue decreases linearly with increasing velocity up to the maximum prolonged speed (U_{mp}) where the endurance is reduced to a fraction of a minute. Surprisingly, extrapolation of this line to fatigue times in the order of a few seconds does not provide a realistic estimate of the absolute maximum burst speed. In this case the extrapolation would predict a maximum burst speed of about 5 $L\,s^{-1}$, whereas the real maximum burst speed is in the order of 7 $L\,s^{-1}$ or may be even higher than that. The logarithm of the endurance time of burst speeds decreases less rapidly than that for prolonged speeds.

Endurance curves have been made from fish induced to swim in flume tanks, fish wheels and in a large annular tank in the Marine Laboratory in Aberdeen. Section 5.3 (page 100) offers general descriptions of these methods. The testing procedure can strongly influence the end result and it is therefore necessary to describe the methods used in more detail here.

Measuring endurance

He and Wardle (1986, 1988) used the annular tank and controlled the speed of mackerel, herring and saithe, using the optomotor reflex. A gantry across the radius of the circular tank can move at angular velocities of 0–1 rad s^{-1}, which correspond to 0–4.5 m s^{-1} at 9 m diameter where the fish were swimming in a 1 m wide annular channel. A slide projector mounted on the gantry casts a light pattern on the tank floor in the channel. Fish follow this pattern and swim at the speed dictated by the moving gantry.

This method offers unconfined swimming space and undisturbed static water conditions. At high velocities, kinematics will be slightly influenced by the effect of continuous turning. In the performance trials, fish were tested either in groups of 2–10 individuals or individually. The velocity of the gantry was increased step by step to the chosen test velocity over a period of 30–60 min. When the velocity reached the predetermined value, the start time for this speed was noted. The behaviour of each individual fish was observed from the rotating gantry. The endurance at each speed was reached when a fish failed to swim with the pattern. Maximum sustained speed (U_{ms}) was defined as the speed with an endurance of more than 200 min. Fish were allowed to rest for 24 h before starting a new endurance test at a different velocity.

Brett (1964) conditioned young sockeye salmon to swim in the respirometer at a speed of about 1–2 $L\,s^{-1}$ for 0.5 h before starting his endurance tests. These fish were taken from large tanks where they had become used to swimming at this velocity over a few weeks. Each endurance test started with 5 min swimming at a velocity of about 3 $L\,s^{-1}$ 'to overcome extreme excitement from more rapid velocity increase'. Subsequently, the velocity was raised in steps to the test level, allowing the fish to adapt during a certain time at each speed level. Fatigue time was reached when the fish could no longer hold itself off the downstream electric screen. U_{ms} was the maximum speed at which the fish could swim for 600 min.

The experimental procedures used in other flume tank endurance experiments differ in details from Brett's (1964) approach. Different speed increment steps and different durations of the intervals between the steps are probably the most significant. Beamish (1978) showed convincingly that these differences are responsible for the fact that there is no consensus in the literature about U_{ms} endurance times. The data, compiled in Table 10.1, are selected from the literature on flume experiments using the criterion that they should allow reconstruction of at least part of the semi-log endurance curve.

Bainbridge (1960) used a transparent fish wheel to record endurance at burst speeds of dace, trout and goldfish. Low, steady speeds could be induced by rotating the wheel backwards, taking the fish away from a striped background. The fish swam to keep station with the background. High burst speeds were evoked using electric shocks of different strengths. The speed obtained was roughly related to the intensity of the shock. The operator rotated the wheel in the direction opposite to the swimming direction. Speed and duration of the rotation were recorded on oscilloscope film. The oscilloscope traces were analysed to find the maximum speeds during arbitrarily selected periods of 1, 2.5, 5, 10, 15 and 20 s.

Compilation of the results

Maximum sustained swimming speeds and log–linear regressions describing the decline of endurance with increasing speed for prolonged and burst swimming speeds are compiled in Table 10.1. The regression equation for the prolonged speeds (U_p in $L\,s^{-1}$) take the form:

$$\text{Log } E_p = aU_p + b \tag{10.1}$$

where E_p is the endurance for prolonged speeds in min, a is the slope of the

Table 10.1 A compilation of sustained, prolonged and burst speed data, collected for fish swimming in the Aberdeen gantry tank (gantry), in a fish wheel (wheel) and in flume tank experiments (flume)

Species	*Temp.* (°C)	*Length* (m)	*Sustained speed*		*Prolonged speed*				*Burst speed*			*Method used*	*Source**
			Max. ($L\ s^{-1}$)	*End.* (*min*)	*logE* = *a*U + *b*		*Max.* ($L\ s^{-1}$)	*End.* (*min*)	*logE* = *c*U + *d*		*Max.* ($L\ s^{-1}$)		
					a	*b*			*c*	*d*			
Pollachius virens	14	0.50	2.2	200	−1.63	5.60	3.4	1.6			6.5	Gantry	1
	14	0.43	2.5	200	−1.52	5.91	3.6	1.6			7.0	Gantry	1
	14	0.35	3.1	200	−1.36	6.16	4.1	3.0			7.6	Gantry	1
	14	0.25	3.4	200	−1.17	5.95	4.9	2.0			8.7	Gantry	1
Scomber scombrus	12	0.31	3.5	200	−0.96	5.45	4.5	15.0			18.0	Gantry	1,2
Clupea harengus	14	0.25	4.1	200	−1.43	8.37	5.5	3.0				Gantry	1
Oncorhynchus mykiss	15±3	0.28							−0.17	−0.22	9.6	Wheel	3
	15±3	0.20							−0.18	−0.07	8.7	Wheel	3
	15±3	0.15							−0.13	−0.25	11.6	Wheel	3
	15±3	0.10							−0.16	−0.10	10.2	Wheel	3
Leuciscus leuciscus	15±3	0.21							−0.18	0.24	11.2	Wheel	3
	15±3	0.20							−0.15	0.01	11.3	Wheel	3
	15±3	0.17							−0.17	0.22	12.0	Wheel	3
	15±3	0.15							−0.16	0.30	11.9	Wheel	3
	15±3	0.14							−0.17	0.18	11.0	Wheel	3
	15±3	0.10							−0.18	0.15	11.0	Wheel	3
	15±3	0.10							−0.14	0.09	13.2	Wheel	3
Carassius auratus	15±3	0.21							−0.23	0.37	9.2	Wheel	3
	15±3	0.16							−0.16	0.25	11.7	Wheel	3
	15±3	0.15							−0.27	0.90	9.3	Wheel	3
	15±3	0.13							−0.34	1.42	9.5	Wheel	3
	15±3	0.12							−0.38	1.74	8.7	Wheel	3
	15±3	0.10							−0.24	1.28	11.6	Wheel	3
	15±3	0.09							−0.29	1.44	11.2	Wheel	3
	15±3	0.07							−0.27	1.21	11.1	Wheel	3
Carassius carassius	13	0.10	3.5	60	−0.36	2.78	11.4	0.1				Flume	4

Oncorhynchus mykiss	17	0.17	4.6	60	−0.91	6.11	7.6	0.1				Flume	4
Cyprinus carpio	15	0.15	4.6	60	−0.81	5.46	7.8	0.2				Flume	4
Oncorhynchus gorbuscha	20	0.55	3.3	600	−1.15	4.97	4.1	1.3				Flume	5
Oncorhynchus nerka	18	0.54	3.1	600	−0.97	4.93	4.4	3.5				Flume	5,7
	10	0.18	2.8	600	−2.10	8.82	4.4	0.3			7.0	Flume	6
Sardinops sagax	19	0.06	3.3	110	−0.38	3.32	11.7	0.1				Flume	8
Odontestes regia	19	0.10	2.6	165	−0.63	3.84	7.6	0.1				Flume	8
Scomber japonicus	19	0.10	2.8	450	−0.62	4.38	8.0	0.3				Flume	8
Anguilla anguilla	12	0.07	3.5	3	−0.42	1.94	5.3	0.7	−0.16	0.63	7.5	Flume	9
Cymatogaster aggregata	15–20	0.09	4.8	13	−0.18	1.95	9.7	1.6			10.9	Flume	10
Hypsurus caryi	15–20	0.14	3.3	13	−0.17	1.65	6.5	3.4				Flume	10
Hyperprosopon argenteum	15–20	0.14	3.3	38	−0.36	2.73	5.8	4.5				Flume	10
Phanerodon furcatus	15–20	0.15	2.9	51	−0.49	3.09	5.2	3.8				Flume	10
Embiotoca jacksoni	15–20	0.15	4.0	10	−0.28	2.11	6.7	1.7			7.5	Flume	10
Chromis punctipinnis	15–20	0.09	7.1	15	−0.14	2.09	12.9	2.4			12.9	Flume	10
Sebastes serranoides	15–20	0.20	3.0	13	−0.28	1.94	5.5	2.5				Flume	10
Genyomemus lineatus	15–20	0.19	4.1	13	−0.37	2.65	5.7	3.2			5.9	Flume	10
Myoxocephalus octodecimspinosus	10	0.20	2.8	7	−0.35	1.76	6.1	0.5				Flume	11
Gadus morhua	8	0.35	2.6	240	−1.13	4.96	3.8	5.6				Flume	11
	5	0.35	2.1	240	−0.99	3.99	3.4	4.2				Flume	11
Sebastes marinus	11	0.16	3.0	240	−0.42	2.94	7.8	0.6				Flume	11
	8	0.16	3.3	13	−0.23	1.70	7.9	0.8				Flume	11
	5	0.17	3.0	14	−0.25	1.71	8.0	0.7				Flume	11
Coregonus clupeaformis	12	0.13	4.1	75	−1.01	5.99	5.8	1.2				Flume	12
	12	0.34	1.7	75	−1.36	4.25	3.0	1.6				Flume	12
	5	0.34	1.4	75	−1.64	4.16	2.2	3.9				Flume	12
Gasterosteus aculeatus	20	0.05	4.0	480	−0.91	6.3	7	1				Flume	13

*Sources: 1, He and Wardle (1988); 2, Wardle and He (1988); 3, Bainbridge (1960); 4, Tsukamoto *et al.* (1975); 5, Brett (1982); 6, Brett (1964); 7, Brett (1967); 8, Beamish (1984); 9, McCleave (1980); 10, Dorn *et al.* (1979); 11, Beamish (1966); 12, Bernatchez and Dodson (1985); 13, Whoriskey and Wootton (1987).

descending line and b is a constant. In the equation for the endurance during burst speeds c and d are used instead of a and b, representing the slope and the constant respectively. The slopes are negative, steeper slopes with the highest negative figures represent the narrowest ranges of swimming speeds; in other words the endurance drops rapidly with minute increases in speed in the most extreme cases. Shallower endurance slopes offer a wider speed range and a more gradual decrease in endurance.

It is difficult to compare the endurance data objectively because of the large variations among the experimental conditions. The temperatures vary between 5 and 20 °C and have not been accurately measured in some of the experiments. The three fundamentally different experimental techniques also complicate comparisons. Slightly different flume tanks or experimental procedures could also obscure general trends. Large deviations among the endurance of different species as well as among individuals of different sizes within a species can be expected. Even individual differences can be as large as 25% (Bainbridge, 1960; Brett, 1964, 1982).

Despite these compatibility and variability problems, this data set can probably be used to show some general trends in maximum sustained and prolonged speeds and in endurance between species. Data from two species indicate the effect of size. Little can be concluded about the effect of temperature owing to inconsistent results.

Values of U_{ms}, U_{mp} and endurance

The U_{ms} of saithe, mackerel and herring in the gantry tank experiments is 3.1 $L s^{-1}$ which is only slightly lower than the average U_{ms} for all the data, including the flume tank experiments, of 3.3 $L s^{-1}$. The maximum prolonged speed (U_{mp}) in the gantry however averages 4.3 $L s^{-1}$, which is substantially lower than the overall average of 6.2 $L s^{-1}$. The average slope of the prolonged speed curves, measured in the gantry tank, is -1.35. This shows that the slopes are much steeper than the overall average value of -0.79. These differences may indicate that fish in the gantry tank were not pushed to their limits. On the other hand, there is a lot of variation among species, even when similar methods are used.

The maximum sustained and prolonged speeds of herring are higher than those of saithe of the same size and the endurance curve is steeper. U_{ms} of mackerel is slightly higher than that of saithe and lower than the value for herring. The endurance of mackerel decreases less rapidly than that of saithe and herring. *Chromis punctipinnis* showed the highest U_{ms} of 7.1 $L\ s^{-1}$ the highest value of U_{mp} of 12.9 $L s^{-1}$. The shallowest endurance slope of -0.14 connects these two points. The U_{ms} of the other species varies between 1.4 and 4.8 $L s^{-1}$. The variation among the U_{mp}s is more evenly distributed between 2.2 and 12.9 $L s^{-1}$. No significant correlation exists between body length L in m and slope a, nor between L and U_{ms} and U_{mp} expressed in $L\,s^{-1}$. Three-quarters of the variation in U_{ms}-values can be explained by the equation:

$$U_{ms} = 0.15 + 2.4\,L \qquad (m\,s^{-1})(r^2 = 0.76,\ N = 33) \qquad (10.2)$$

when U_{ms} is in $m\,s^{-1}$. Only 64% of the variation in U_{mp}s (in $m\,s^{-1}$) is explained by:

$$U_{mp} = 0.61 + 2.4\,L \qquad (m\,s^{-1})(r^2 = 0.64,\ N = 33) \qquad (10.3)$$

These correlations only reflect the fact that larger fish can swim at higher absolute speeds. Note that the slopes in Equations 10.2 and 10.3 are equal despite the wide variety of endurance times at which both velocities, but in particular U_{mp}, have been determined. The lines representing the allometric Equations 10.2 and 10.3 together with the data points are depicted in Fig. 10.1. The difference between U_{ms} and U_{mp} is constant at 0.5 $m\,s^{-1}$.

The steepness of the endurance curves (Equation 10.1) for prolonged speeds varies considerably among species. The most extreme value of −2.1 was found by Brett (1964) for young sockeye salmon. Brett's (1967, 1982) results with adults of sockeye and pink salmon, *Oncorhynchus gorbuscha*, are −0.97 and −1.15 respectively. These figures are of the same order of magnitude as those of cod, trout, lake whitefish, saithe, mackerel

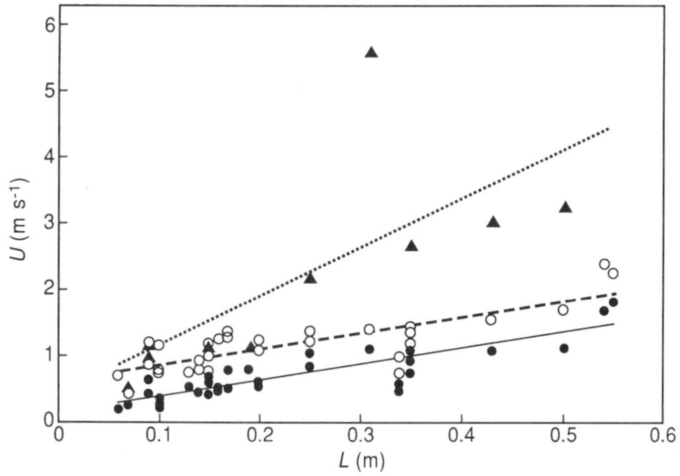

Fig. 10.1 Maximum sustained (U_{ms}), maximum prolonged (U_{mp}) and maximum burst (U_{mb}) speeds in $m\,s^{-1}$ as a function of body length (L) in m. Based on data shown in Table 10.1. Symbols: (●), U_{ms}; (○), U_{mp}; (▲), U_{mb}. Curves: solid curve, U_{ms}, Equation 10.2; broken curve, U_{mp}, Equation 10.3; stippled curve, U_{mb}, Equation 10.4.

and herring. Interspecific differences in body size and temperature do not seem to have much influence on the approximate level of the slope value. Carp, Pacific mackerel and *Odontestes regia*, have intermediate a values. The slopes of all the other species are very shallow. The most extreme example is, as mentioned before, *Chromis punctipinnis*. This 8.5 cm animal had a wide range of endured swimming speeds, starting with a U_{ms} of 7.1 $L\,s^{-1}$ with a measured endurance of 15 min up to a U_{mp} of 12.9 $L\,s^{-1}$ which it could keep for 2.4 min. Many of the species with shallow endurance curves have some features in common: they are usually small demersal species living in complex (mostly coastal) environments and are not pelagic, long-distance swimmers. Most of these use the pectoral fins for sculling at low speeds and will only swim with movements of body and tail at higher velocities; *Cymatogaster* and *Embiotoca* use their pectoral fins also at higher speeds.

In some species, for example the Pacific sardine, *Sardinops sagax*, the semi-log prolonged speed curve remains linear up to very small endurance values (of 0.05 min in the example). These species do not seem to have a separate burst speed endurance curve.

Surfperches (Embiotocidae) and sticklebacks (Gasterosteidae) belong to the few groups of fish that employ pectoral fin propulsion at high sustained swimming speeds. *Cymatogaster* swims for 45 min at 4 $L\,s^{-1}$ (Webb, 1973) and can keep up 4.8 $L\,s^{-1}$ for 13 min (Dorn *et al.*, 1979) rowing with its pectorals.

The endurance of three-spine sticklebacks, involving males of 4.4–5.2 cm and females of 4.8–5.3 cm, was tested in a flume at 20 °C by Whoriskey and Wootton (1987). Most animals were able to cover more than 6 km in 8 h swimming at 4 $L\,s^{-1}$. At 5–6 $L\,s^{-1}$ a few fish were still using the pectoral fins and managed to do that for 8 h. Individual differences were large at these intermediate prolonged speeds; the distances covered by individual males, for example, varied between 75 and 7200 m before they fatigued. At 7 $L\,s^{-1}$ all fish would use body and tail locomotion and the endurance of all animals dropped abruptly to 1 min or less.

Effect of size and temperature

The data for saithe and *Coregonus* offer the opportunity to study size effects more precisely within a single species. In saithe, U_{ms} in $L\,s^{-1}$ is proportional to $L^{-0.63}$ and the $Q_{10\,cm}$ value is 0.83 ($Q_{10\,cm}$ is here the rate of speed change in $L\,s^{-1}$ for every 10 cm length change). He (1986) found maximum speeds of saithe at 12 °C to be proportional to $L^{-0.43}$ ($L\,s^{-1}$), representing a $Q_{10\,cm}$ value of 0.88. This figure is virtually the same as the $Q_{10\,cm}$ values found for the maximum tail beat frequencies in Table 7.2. The slopes of the prolonged speeds regression curves of saithe (Table 10.1) become steeper with increasing body length. The slope values (a) decrease linearly with body length at a rate of 0.18 for every 10 cm length increase. The $Q_{10\,cm}$ value for the slope change with length is 1.14. The endurance curves for saithe of different lengths are drawn as an example in Fig. 10.2. These curves tell us that a 0.25 m fish can swim at 3 $L\,s^{-1}$ continuously without fatigue but that a fish twice that size can swim for only 5 min at the same relative speed. A 0.5 m saithe can keep up a speed of 1 $m\,s^{-1}$ indefinitely, whereas a 0.25 m fish would be exhausted after 19 min. (Equation 10.2 over-estimates the U_{ms} values for the largest saithe and predicts a value lower than measured for the 0.25 m saithe.)

Coregonus of 0.34 and 0.13 m were measured at the same temperature. The $Q_{10\,cm}$ values for U_{ms} and U_{mp}, for a 1.6 min endurance, were 0.67 and 0.74 respectively. This means that U_{ms} and Ur_{mp} decrease more rapidly with increasing size in *Coregonus* than in saithe. The slope of the prolonged speed curve steepens with increasing length with the same factor as found for saithe.

For cod, redfish, *Sebastes marinus*, and *Coregonus*, results of measurements at different temperatures of animals of the same length have been published. The trends emerging from these few data are not consistent. A decrease in temperature coincides with a decrease in U_{ms} and U_{mp} in cod and *Coregonus*, but not in *S. marinus*. The slope values of cod and *S. marinus* increase (the

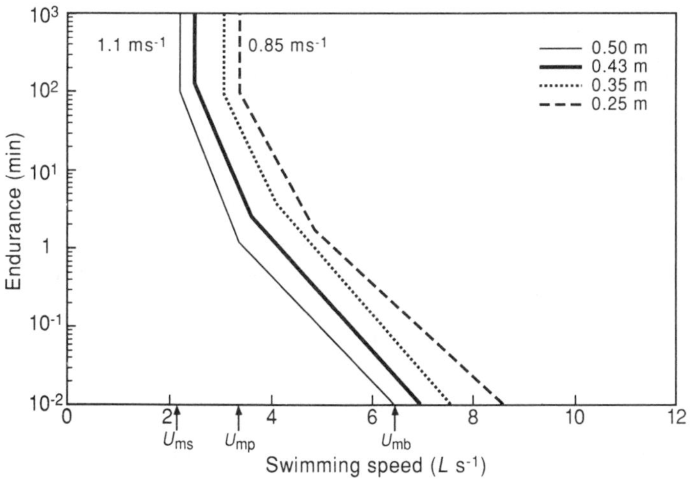

Fig. 10.2 The effect of size on endurance of saithe, spontaneously following light patterns at increasing speeds. Endurance curves are shown for four individual fish. Maximum sustained swimming speed (U_{ms}), maximum prolonged speed (U_{mp}) and maximum burst speed (U_{mb}) are indicated for the largest (0.5 m) fish. Values of U_{ms} range from 0.85 $m\,s^{-1}$ for the smallest fish to 1.1 $m\,s^{-1}$ for the largest.

slopes become shallower) and the slope of *Coregonus* is shown to be steeper after a drop in temperature.

The effects of schooling and genetic differentiation

Many of the endurance tests have been conducted with more than one fish at a time. Many species tend to be more quiescent when they are in a small school. The influence of schooling on the U_{ms} was investigated by Kawamura *et al.* (1988). Small schools were used of three individuals each of killifish, *Oryzias latipes*, tilapia, *Tilapia nilotica*, and mullet, *Mugil cephalus*. Each trio consisted of animals of slightly different sizes. The U_{ms} of the schools were usually close to that of the figure found for the fastest individual when tested individually. Hence, the other two members were swimming faster than they would do on their own.

There are also genetically determined intraspecific differences in endurance. Taylor and Foote (1991) tested two forms of *Oncorhynchus nerka*: the anadromous sockeye salmon, migrating to the sea until mature at about 5 kg, and the kokanee which stays in fresh water to become mature at a body weight of only 0.5 kg. Hybrids between these groups are common. Fish aged about 6 months from both groups, about 8 cm long, were tested in a flume at 10 °C. The speed of the flow was increased from 2.5 $L\,s^{-1}$ upwards with steps of 1 $L\,s^{-1}$ each hour until all fish were fatigued. The penultimate speed at which the fish could swim for the entire 60 min before fatigue was taken as a measure for endurance. Sockeye salmon reached 8.3 $L\,s^{-1}$ against 7.3 $L\,s^{-1}$ for the kokanee.

Three-spine sticklebacks, *Gasterosteus aculeatus* occur in land-locked and anadromous populations. The anadromous fish have larger pectorals and a more slender body than the animals remaining in fresh water. The morphological differences are reflected in the maximum stride lengths of 0.23 and 0.19 L for sea-going and freshwater fish respectively. At 5 $L\,s^{-1}$ the anadromous fish fatigued less easily, showing a sevenfold endurance of 129 against 18 min for the freshwater fish. Burst swimming of freshwater sticklebacks at 14.9 $L\,s^{-1}$ was significantly faster than the escape speeds of 11 $L\,s^{-1}$ of anadromous fish (Taylor and McPhail, 1986).

Burst speeds

Burst speed curves of trout, dace and goldfish have been determined by Bainbridge (1960), and McCleave (1980) offers one for the eel. The slopes are −0.20 on average, which is almost one-fourth as steep as the average angle of the prolonged speed curves. Bainbridge measured animals of different sizes, but there is no apparent size effect.

The maximum burst speed measured was 10 $L\,s^{-1}$ on average for fish with a mean body length of 18 cm. The maximum burst speed U_{mb} in $m\,s^{-1}$ in relation to body length (m) is best described by:

$$U_{mb} = 0.4 + 7.4\,L \qquad (m\,s^{-1})(r^2 = 0.64,\ N = 30) \qquad (10.4)$$

U_{mb} increases less rapidly with length than the average value of 10 $L\,s^{-1}$ predicts. In other words, 10 $L\,s^{-1}$ as a first estimate for the maximum burst speed is reasonably close for fish between 10 and 20 cm body length. Equation 10.4 predicts relative burst speeds for fish larvae of 1 cm of about 47 $L\,s^{-1}$, which is in the order of magnitude reported for larval fish (Kaufmann, 1990). Fuiman (1986) measured the effect of temperature on the burst swimming speeds of 3.6 mm zebra danios. He found startle responses to electric shocks of 66 $L\,s^{-1}$ at 30 °C. The maximum burst swimming speed at 21 °C was 50 $L\,s^{-1}$. Equation 10.4 overestimates the maximum performance of these very small larvae.

The highest speed in Table 10.1 is that of the 0.305 m mackerel swimming at 18 $L\,s^{-1}$. The fish was filmed, swimming in water of 12 °C, using a high speed camera in fixed position, running at 200 frames s^{-1} (Wardle and He, 1988). The mackerel used a tail beat frequency of 18 Hz and a stride length of 1 L. The muscle twitch contraction time at the same temperature for mackerel of the same size was 0.026 s, predicting a maximum tail beat frequency of 19 Hz. The maximum speed predicted on the basis of muscle twitch contraction times of the bluefin tuna, by Wardle *et al.* (1989), was 15 $m\,s^{-1}$ assuming a moderate stride length of 0.65 L at a tail beat frequency of 10 Hz. This value is close to the 17 $m\,s^{-1}$ predicted by Equation (10.4); a stride length of 0.75 L would provide that speed at the same frequency.

The relationships between velocity and body length as shown by Equations 10.2, 10.3 and 10.4 for U_{ms}, U_{mp} and U_{mb} respectively are visualized in Fig. 10.1 together with the data points. The two parallel lines for U_{ms} and U_{mp} indicate that the difference between these speeds is constant at about 0.5 $m\,s^{-1}$. This configuration of the speed–length relationships is highly enigmatic.

Performance and metabolic rate

Brett's (1964) classical results are used to exemplify the relation between metabolic rate and endurance. Fig. 10.3 shows the energetic measurements and the endurance curve in one graph in relation to swimming speed. Both curves are based on data for 18 cm sockeye salmon, weighing 0.05 kg, tested at 10 °C. Table 9.2 contained data for the same fish showing the optimum speed of 1.6 $L\,s^{-1}$ and the *COT* of 0.24 $J\,N^{-1}\,m^{-1}$ at that speed but at a temperature of 15 °C. The sockeye salmon performed maximally at this temperature, but no endurance test results were given. The standard metabolic rate (*SMR*) at 10 °C is at about 0.24 $W\,kg^{-1}$, 0.012 W. The active

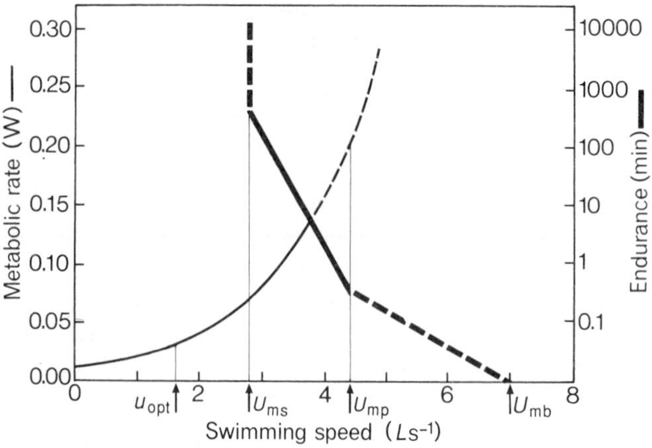

Fig. 10.3 The metabolic rate (linear scale) and the endurance (logarithmic scale) of a 0.18 m, 0.05 kg sockeye salmon as functions of swimming speed in Ls^{-1}. The water temperature was 15 °C. The optimum swimming speed (u_{opt}), the maximum sustained speed (U_{ms}), the maximum prolonged speed (U_{mp}) and an estimate of the maximum burst speed (U_{mb}) are indicated. Based on Brett (1964).

metabolic rate increases with speed to a value of 0.033 W (2.75 times *SMR*) at the optimum speed of 1.6 Ls^{-1} (the optimum speed turns out to be the same for 10 and 15 °C). The maximum speed that can be sustained for 10 h is 2.8 Ls^{-1} and requires 0.072 W (6 *SMR*). At about 3.7 Ls^{-1}, sockeye salmon of this size and at 10 °C use as much oxygen as they can, which is about 630 mg O_2 $kg^{-1}h^{-1}$ (at 15 °C the maximum figures were 988 mg O_2 $kg^{-1}h^{-1}$ swimming at 4.1 Ls^{-1}). This temperature-dependent aerobic limit corresponds with a metabolic rate of 0.128 W (11 *SMR*). The endurance at this speed is still about 5 min. Swimming at higher speeds requires more oxygen than the fish can take in and will cause serious oxygen debt. The endurance at the maximum prolonged speed of 4.4 Ls^{-1} is about 18 s and the fish is using energy at a rate of about 0.2 W, that is if we trust the extrapolated curve for the metabolic costs beyond the maximum oxygen intake value. If that is the case, the maximum burst speed near 7 Ls^{-1} requires almost 90 *SMR* and 32 times the rate of energy use at u_{opt}. The cost of transport at U_{mb} would be about 7.5 times as high as the *COT* at u_{opt}.

10.3 ALLOCATION OF ENERGY

On coral reefs, in clear warm water, the opportunity arises to study the time course of the natural behaviour of an individual fish for hours on end.

Students of my research group made time budgets of herbivorous parrotfishes (Scaridae) over the past 8 years on the leeward reef of the island Bonaire (Netherlands Antilles). We follow the animals underwater using scuba diving equipment. Underwater event recorders with 16 keys are used to record time sequences of components of behaviour (including swimming, feeding, defecating, spawning, agonistic interactions, resting, etc.) at 0.1 s precision.

This approach is not possible in cold temperate waters with low visibility. An alternative method was used by Lucas *et al.* (1991) who followed the behaviour of pike in Scottish lochs for several days using heart rate telemetry.

We expect a purely herbivorous animal to spend more time on feeding activities than a carnivorous predator, because of the large differences in the calorific values of the food source.

Time allocated to swimming

The parrotfishes are feeding mainly on small turfs of algae, on algal crusts and on endolithic algae growing on and in dead coral. The five species we looked at are only active in daytime in shallow water and sleep at night between coral heads on the reef slope. These animals are protogynous hermaphrodites with a complicated sex life; the social structure of the populations is also rather complicated (Robertson and Warner, 1978). Some of the individuals were recorded regularly during more than 4 years.

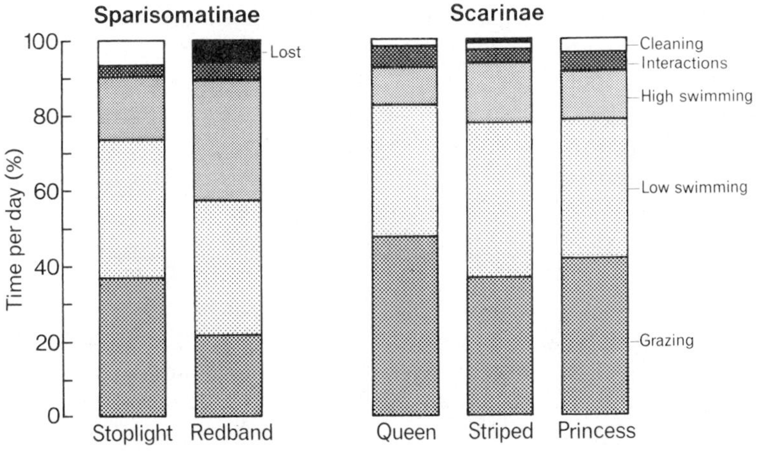

Fig. 10.4 Time budgets of territorial males of five species of Caribbean parrotfish. Average values for the behavioural categories are obtained by collecting one-hour samples of the durations of these categories evenly distributed between dawn and dusk. See text for further explanation.

Fig. 10.4 compares the typical daily time budgets of territorial males of the stoplight and the redband parrotfish (*Sparisoma viride* and *S. aurofrenatum*) belonging to the subfamily of the Sparisomatinae with the time budgets of territorial males of three species of the subfamily Scarinae: the queen, *Scarus vetula*, the striped, *S. iserti*, and the princess, *S. taeniopterus*, parrotfish. The behavioural categories, defined in the text below, are 'grazing', 'low and high swimming', 'interactions', 'cleaning' and 'lost'. The daily behaviour of parrotfishes follows a fixed pattern, with about 90% of the time devoted to feeding. The general pattern is that the fish takes one or two bites at a certain spot (grazing) and swims a few metres to the next feeding spot (low swimming) where it takes another bite. To cover larger distances, the fish swim higher in the water column; this occurs during territorial defence behaviour, spawning activities and while swimming to and from the sleeping sites. The redband spends more time swimming high than the other species because it feeds frequently on planktonic organisms in the water column (that is also the place where the animal gets lost more often than while grazing near the substrate). Most swimming movements are made with the pectoral fins, with occasional fast movements involving undulations of body and tail. These occur mostly during spawning or agonistic behaviour, which are grouped together under the category 'interactions'. Most species devote some time to being cleaned by the cleaning goby, *Gobiosoma genie*. The general picture emerging from Fig 10.4 is that, despite small differences, parrotfish spend 50% or more of the daytime swimming. This amounts to more than 6 h per 24 h. Swimming using tail beats varies between about 3% of the time for the stoplight territorial male to 5% of the time budget of the queen parrotfish. Virtually all swimming activities are in connection with either feeding or reproduction. Swimming to avoid predation rarely occurs and has only occasionally been observed.

Lucas *et al.* (1991) fitted three pike with a heart beat transmitter and released them near the spot where they were caught. One was subsequently tracked for about 72 h and a second one for 90 h. The third fish was followed for only 20 h. Locomotory activity was associated with heart rate elevations in combination with EMG interference signals. Signals were sampled every 5 min. The number of 5 min periods showing swimming activity was added up to obtain an estimate of the total time allocated to swimming. Swimming of pike occurred during the day as well as during darkness. The first pike was the most active one, swimming 5.8 h per 24 h on average. The second pike swam 3.5 h per natural day. The third fish was digesting a perch on release and did not show a lot of activity during the 20 h tracking period. Pike is usually regarded as an ambush predator which only swims for very short periods. These tracking results in the wild change that image. Many more time budgets of fish in their natural environment are needed to reveal the real general trends.

Activity as part of the energy budget

Time budgets form the logical basis for energy budgets but transformation from one to the other is not straightforward. An energy budget is usually represented by an equation such as:

$$C = P + R + E \tag{10.5}$$

where C represents the income side, being the energy consumed. Of the expenses on the right-hand side of the equation, E stands for the energy content of excretory products such as faeces and urine, P for the metabolic costs of production and R for the energy used by respiration. All these factors can be represented in joules, the units of energy (Calow, 1985). The diagram of Fig. 10.5 schematically represents the balance sheet. The food consumed is assimilated into metabolizable energy. Faeces and urine are the waste products of this process. Assimilation, transport and storage of the useful fraction requires energy commonly indicated as the specific dynamic action (SDA) or 'heat increment', a fraction of the total respiration. Some of the metabolizable energy is burnt to accommodate the costs of resting or standard metabolism. A larger fraction is required to fuel the active metabolism

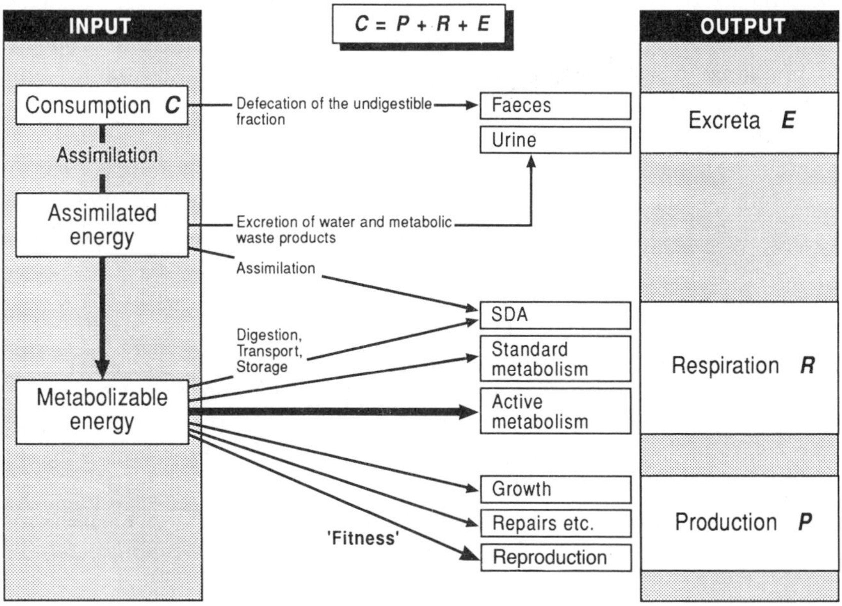

Fig. 10.5 A generalized energy balance sheet for fish showing the main components and allocation of energy obtained by ingesting food.

including locomotory activities. All the fractions of R are highly variable, depending on a variety of factors among which temperature and swimming speed are dominant. The production P also consists of a number of variable fractions such as growth, the production of slime, damage-repair and reproduction. Reproduction can be regarded as the balancing item on the sheet. Energy allocated to the production of gonads represents, in energetic terms, the Darwinian fitness of an individual. The relation between the energy allocated to swimming and any of the other components of the balance sheet is complex, and only a limited number of studies reveals something of the nature of this complexity.

Swimming and the acquisition of food

Several studies have shown that food intake (C) and swimming velocities are closely linked. Robinson and Pitcher (1989) showed how routine speeds of herring larvae depend on the hunger state and can change at the instant of feeding. Routine swimming speeds of food-deprived fish were about 1.5–2 $L\,s^{-1}$, which was substantially lower than the average velocities between 2.5 and 3 $L\,s^{-1}$ of satiated herring. After feeding, swimming speeds increased a little in satiated fish and doubled in the hungry fish. Under both circumstances, speed gradually decreased after about 5 min after food was provided. The Cape anchovy, *Engraulis capensis*, employs specific swimming behaviour and speeds for different food items, probably to maximize food intake and minimize energy expenditure (James and Findlay, 1989). During normal swimming without feeding these animals use 4–5 tail beats followed by a glide. Speeds vary between 0.7 and 1.9 $L\,s^{-1}$. During feeding on phytoplankton a filter-feeding technique is used. A filtering bout starts with opening of the mouth and the opercula followed by 3–12 strong tail beats with a large amplitude. It ends after between 0.4 and 3 s with closure of the mouth just before the end of the last tail beat. Swimming speeds and turning rates are not different from routine swimming in the absence of food. There is no orientation towards the food particles. Swimming speeds during filter feeding on microzooplankton were faster at about 2.1 $L\,s^{-1}$. Turning rates were higher than measured during filtering of phytoplankton and positively correlated with the concentration of food. Filter feeding was accompanied by dense schooling. Schools broke down immediately after the introduction of particulate food. Fish start to hunt individually usually after an initial feeding frenzy. Hunting for particles is done at high speeds of 2.4 $L\,s^{-1}$. The velocity appears to be independent of the food concentration, in contrast to the turning rates which increase significantly with higher concentrations of food items. The fish use large-amplitude tail beats interspersed with sporadic short glides preceding strikes. There is no difference in behaviour between light and dark. The respiration

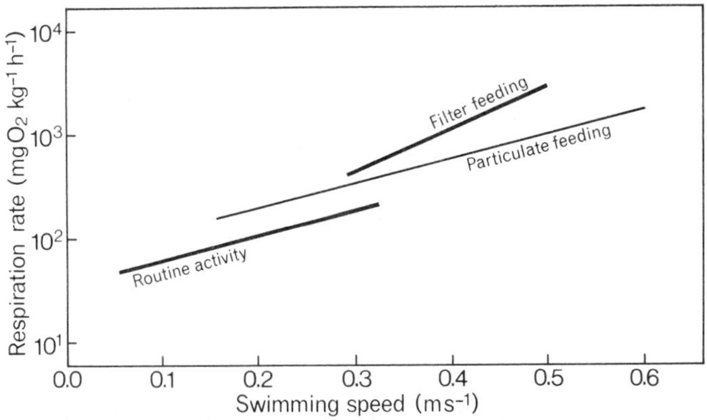

Fig. 10.6 Respiration rates at the swimming speeds used during routine swimming, filter-feeding and particulate feeding of the Cape anchovy, *Engraulis capensis*, average length 9 cm, at 16 °C. Redrawn from James and Probyn (1989).

rates during routine swimming, filter-feeding and particulate feeding are different. Fig. 10.6 shows that the slope of the relation between filter-feeding respiration and speed was steeper than for particulate and routine swimming. Filter-feeding is obviously more expensive owing to increased drag caused by the open mouth and opercula (James and Probyn, 1989).

The anchovy example showed that fish may invest energy differently to obtain different food sources. However the net benefit from feeding on each of these sources has not, to my knowledge, been investigated. There are, as is the case in mammals, reciprocal relationships between diet and performance in fish. The influence of the diet on maximum performance has been tested for the lake trout, *Salvelinus namaycush*, by Beamish *et al.* (1989). Three diets of approximately the same caloric value of about 21 $MJ\,kg^{-1}$ were fed to three groups of juvenile lake trout during 70 days. The diets varied in protein and lipid content. One was high in lipid and low in protein, another low in lipid and high in protein, and a third one had intermediate values. The maximum swimming speeds the fish could endure during 60 min in a flume at 10 °C increased between day 0 and day 70 for all three diets, but less so for the lipid-rich and intermediate diets than for the protein-rich diet. The lipid-rich diet yielded an increase from 6.3 to 6.6 $L\,s^{-1}$ and the protein-rich diet an increase from 6.2 to 7.4 $L\,s^{-1}$. Resting metabolism did not change with diet.

Speed, growth and reproduction

The effect of exercise on growth rates is of special interest for fish farmers who, like all farmers, want the highest food conversion and fastest growth of their animal product. East and Magnan (1987) investigated the relation between sustained swimming speeds and growth for the brook charr, *Salvelinus fontinalis*, at 17 °C. Four groups of 25 fish of about the same length were kept in four containers, each with its own water velocity of 0, 0.85, 1.72 and 2.5 $L\,s^{-1}$ (where L is the average fish length in each container). After 20 days the length increases were on average 1, 3.5, 0.5 and 0.7% for the groups swimming at 0, 0.85, 1.72 and 2.5 $L\,s^{-1}$ respectively.

Weight increased for the groups in the same order by 24, 34, 16 and 17%. The changes in length and the increases in weight, apart from the weight increase of the 0.85 $L\,s^{-1}$ group, are not significant. But there is a clear trend: the largest values are found for the swimming speed of 0.85 $L\,s^{-1}$, and there are consistently lower values at the higher speeds. Davison and Goldspink (1977) did similar tests with *Salmo trutta* and found maximal growth at 1 $L\,s^{-1}$, and Greer Walker and Emerson (1978) found an optimum growth speed of 1.5 $L\,s^{-1}$ for *Oncorhynchus mykiss*. It is not easy to explain the mechanism behind the increased growth at low speeds instead of at no swimming. The behaviour of the animals is different when they swim in a flow instead of standing water. Schooling becomes unidirectional and more dense with increased flow. There were three times as many aggressive interactions at zero velocity than while swimming at intermediate speeds. The diminished growth rates at high speeds are easier to explain by pointing to the fact that high-speed swimming requires energy that cannot be allocated to growth.

Koch and Wieser (1983) contributed to the idea of energy partitioning or trade-off between activity and another function with a study on the relation between swimming activity, oxygen consumption and growth of gonads of the roach, *Rutilus rutilus*. During one seasonal cycle the swimming activity is reduced in the period of gonad synthesis. Reduced locomotion saves 1485 $kJ\,kg^{-1}$, and gonadal tissue with an energy content of 364 $kJ\,kg^{-1}$ is synthesized. This trade-off is depicted in Fig. 10.7. Spawning takes place in June and new gonads are formed from July to December. It is easy to see that it is not a temperature effect because the activity in August–September–October is much lower, at temperatures decreasing from 18 to 10 °C, than during April–May–June when the swimming activity is higher under the same temperature conditions. The inset of Fig. 10.7 nicely demonstrates the trade-off. Both temperature and activity increase from December to June, where the increased activity is precisely matched by an increase in oxygen consumption. In the second half of the season, activity drops faster than can be explained by the temperature effect alone. Each temperature allows a certain maximum oxygen uptake used for activity needed to collect food for maintenance, growth and reproduction. In the first half of the year all the energy collected goes to maintenance and growth. In the second half of the year there is only one option: keep oxygen uptake maximal and use it for

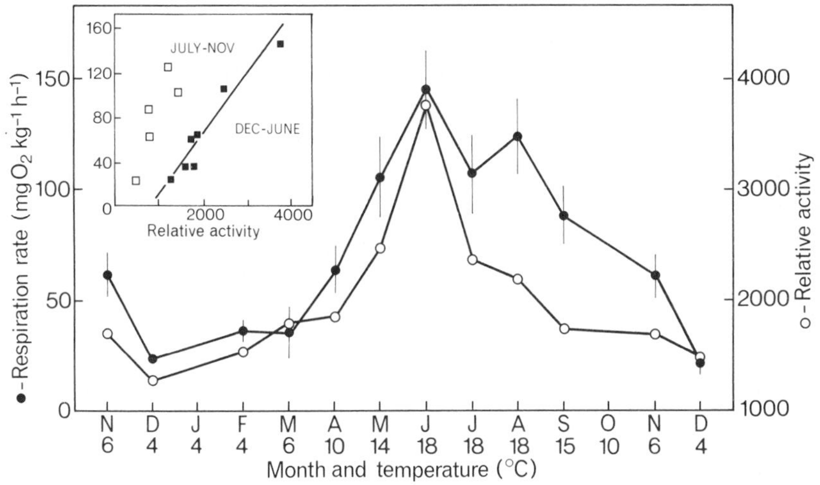

Fig. 10.7 Seasonal cycle of oxygen consumption and relative activity of roach. Measurements were made at the temperatures indicated below the abscissa, prevailing in the natural habitat in each month. Oxygen consumption is plotted against relative activity in the inset. Reproduced with permission from Koch and Wieser (1983).

both synthesis of gonads and activity. This indicates that oxygen uptake is the limiting factor.

10.4 SUMMARY AND CONCLUSIONS

A comparison of published data reveals that the average maximum sustained speed for fish, varying in size between 5 and 54 cm, is in the order of 3 $L\,s^{-1}$ and that the speed at which fish are able to swim during at least a fraction of a minute, is about twice that value. The average slope of the endurance curve is close to that found for carp. A 15 cm representative of this species was capable of swimming at 4.6 $L\,s^{-1}$ for 60 min and at 7.8 $L\,s^{-1}$ for 0.2 min. Significant linear correlations are found between body length and maximum sustained and prolonged speeds expressed in $m\,s^{-1}$. The slopes of these lines are identical, and hence the difference between these two speeds is constant at 0.5 $m\,s^{-1}$. Within a single species, maximum sustained and prolonged speeds expressed in $L\,s^{-1}$ line with increasing length and so does endurance.

Demersal fish living in complex environments have shallower endurance curves than pelagic long-distance swimmers which fatigue more quickly when they break the limit of the maximum sustained speed.

The maximum burst speed in $m\,s^{-1}$ increases linearly with body length; the slope is about three times as steep as that of the maximum sustained and prolonged speeds. The average relative value is about 10 $L\,s^{-1}$, a figure that turns out to be a fairly good estimate for fish between 10 and 20 cm long. Small fish larvae swim at up to 60 $L\,s^{-1}$ during startle response bursts.

Endurance in fish swimming at prolonged speeds is limited by the oxygen uptake capacity. Higher speeds cause serious oxygen debts.

Both herbivorous and carnivorous fish may spend a substantial part of the time swimming, either to obtain food or in relation to reproductive activities. Swimming to escape from predators is of utmost importance for survival but is an insignificant portion of the time budget.

The components of an energy budget show all kinds of reciprocal relationships. The studies selected to exemplify such relationships lead to the following conclusions.

1. Hungry fish swim slower than satiated fish during routine swimming and faster after feeding started.
2. There is circumstantial evidence that a fish may choose different swimming strategies for different food sources to optimize the net energy gain.
3. A protein-rich diet may increase maximum sustained swimming speeds, more so than a lipid-rich diet.
4. Moderate continuous swimming speeds provide the best conversion of food into growth if compared with faster swimming fish and fish that had no exercise at all.
5. Fish may show a trade-off between swimming activity and the production of gonads.

References

Addink, A.D.F., Waversveld, F. van, Thillart G. van den, Boon J. van der, Lugtenburg, J. and Waarde, A. van, (1989) Direct and indirect calorimetry and nuclear magnetic resonance spectroscopy on intact fish during normoxia and anoxia, in *Energy Transformations in Cells and Organisms* (eds W. Wieser and E. Gnaiger), Georg Thieme, Stuttgart, pp. 140–44.

Akster, H.A. (1981) Ultrastructure of muscle fibres in head and axial muscles of the perch (*Perca fluviatilis* L.): a quantitative study. *Cell Tissue Res.*, **219**, 111–31.

Akster, H.A. (1985) Morphometry of muscle fibre types in the carp (*Cyprinus carpio* L.): relationships between structural and contractile characteristics. *Cell Tissue Res.*, **241**, 193–201.

Akster, H.A., Granzier, H.L.M. and Focant, B. (1989) Differences in I band structure, sarcomere extensibility, and electrophoresis of titin between two muscle fibre types of the perch (*Perca fluviatilis*). *J. Ultrastruct. Mol. Struct.* Res., **102**, 109–21.

Akster, H.A., Granzier, H.L.M. and Keurs, H.E.D.J., te, (1985a) A comparison of quantitative ultrastructural and contractile characteristics of muscle fibre types of the perch, *Perca fluviatilis* L. *J. comp. Physiol.*, **155B**, 685–91.

Akster, H.A., Granzier, H.L.M., Osse, J.W.M. and Terlouw, A. (1985b) Muscle fibre types and muscle function in fish, in *Fortschritte der Zoologie/Vertebrate morphology* (eds H.R. Duncker and E. Fleischer), Gustav Fischer, Stuttgart, pp. 27–30.

Alexander, R.McN. (1965) The lift produced by the heterocercal tail of selachii. *J. exp. Biol.*, **43**, 131–8.

Alexander, R.McN. (1969) Orientation of muscle fibres in the myomeres of fishes. *J. Mar. Biol. Ass. U.K.*, **49**, 263–90.

Alexander, R.McN. (1981) *The Chordates*, Cambridge University Press, Cambridge, 510 pp.

Alexander, R.McN. (1988) *Elastic Mechanisms in Animal Movement*, Cambridge University press, Cambridge, 141 pp.

Aleyev, Y.G. (1969) *Function and Gross Morphology in Fish* (Transl. M.Raveh, ed. H. Mills), Keter Press, Jerusalem, 268 pp.

Aleyev, Y.G. (1977) *Nekton*, Dr W. Junk, The Hague, 435 pp.

Altringham, J.D. and Johnston, I.A. (1989) The innervation pattern of fast myotomal muscle in the teleost *Myoxocephalus scorpius*: a reappraisal. *Fish Physiol. Biochem.*, **6**, 309–13.

Altringham, J.D. and Johnston, I.A.(1990a) Modelling muscle power output in a swimming fish. *J. exp. Biol.*, **148**, 395–402.

Altringham, J.D. and Johnston, I.A. (1990b) Scaling effects on muscle function: power output of isolated fish muscle fibres performing oscillatory work. *J. exp. Biol.*, **151**, 453–67.

Altringham, J.D., Wardle, C.S. and Smith, C.I. (1993) Myotomal muscle function at different points on the body of a swimming fish. *J. exp. Biol.*, **182**, in press.

Anderson, M.E. and Johnston, I.A. (1992) Scaling of power output in fast muscle fibres of the atlantic cod during cyclical contractions. *J. exp. Biol.*, **170**, 143–54.

Archer, S.D., Altringham, J.D. and Johnston, I.A. (1990) Scaling effects on the neuromuscular system, twitch kinetics and morphometrics of the cod, *Gadus morhua*. *Mar. Behav. Physiol.*, **17**, 137–46.

Archer, S.D. and Johnston, I.A. (1989) Kinematics of labriform and subcarangiform swimming in the antarctic fish *Notothenia neglecta*. *J. exp. Biol.*, **143**, 195–210.

Arita, G.S. (1971) A re-examination of the functional morphology of the soft-rays in teleosts. *Copeia*, **1971**, 691–7.

Arnold, G.P. and Weihs, D. (1978) The hydrodynamics of rheotaxis in the plaice (*Pleuronectes platessa*). *J. exp. Biol.*, **75**, 147–69.

Baerends, G.P. and Baerends-van Roon, J.M. (1950) An introduction to the study of the ethology of cichlid fishes. *Behaviour*, Supp. 1, 1–242.

Bainbridge, R. (1958a) The speed of swimming of fish as related to size and to the frequency and amplitude of the tail beat. *J. exp. Biol.*, **35**, 109–33.

Bainbridge, R. (1958b) The locomotion of fish. *New Scient.*, **24**, 476–8.

Bainbridge, R. (1960) Speed and stamina in three fish. *J. exp. Biol.*, **37**, 129–53.

Bainbridge, R. (1963) Caudal fin and body movement in the propulsion of some fish. *J. exp. Biol.*, **40**, 23–56.

Bainbridge, R. and Brown, R.H.J. (1958) An apparatus for the study of the locomotion of fish. *J. exp. Biol.*, **35**, 134–7.

Batty, R.S. (1981) Locomotion of plaice larvae, in *Vertebrate Locomotion*, (Symposia of the Zoological Society of London) (ed. M.H. Day), Academic Press, London, pp. 53–69.

Batty, R.S. (1984) Development of swimming movements and musculature of larval herring (*Clupea harengus*). *J. exp. Biol.*, **110**, 217–29.

Batty, R.S. and Blaxter, J.H.S. (1992) The effect of temperature on the burst swimming performance of fish larvae. *J. exp. Biol.*, **170**, 187–201.

Baudinette, R.V. and Gill, P. (1985) The energetics of 'flying' and 'paddling' in water: locomotion in penguins and ducks. *J. comp. Physiol.*, **155B**, 373–80.

Beamish, F.W.H. (1966) Swimming endurance of some Northwest Atlantic fishes. *J. Fish. Res. Bd. Can.*, **23**, 341–7.

Beamish, F.W.H. (1970) Oxygen-consumption of largemouth bass, *Micropterus salmoides*, in relation to swimming speed and temperature. *Can. J. Zool.*, **48**, 1221–8.

Beamish, F.W.H. (1978) Swimming Capacity, in *Fish Physiology*, Vol. VII, *Locomotion* (eds W.S. Hoar and D.J. Randall), Academic Press, London, pp. 101–87.

Beamish, F.W.H. (1984) Swimming performance of three Southwest Pacific fishes. *Mar. Biol.*, **79**, 311–13.

Beamish, F.W.H., Howlett, J.C. and Medland, T.E. (1989) Impact of diet on metabolism and swimming performance in juvenile lake trout, *Salvelinus namaycush*. *Can. J. Fish. Aquat. Sci.*, **46**, 384–8.

Becerra, J., Montes, G.S., Bexiga, S.R.R. and Junqueira, L.C.U. (1983) Structure of the tailfin in teleosts. *Cell Tissue Res.*, **230**, 127–37.

Berger, M. (1985) Sauerstoffverbrauch von Kolibris (*Colibri coruscans und C.thalassinus*) beim Horizontalflug, in *Biona Report* (ed. W. Nachtigall), G. Fisher, Stuttgart, Akad. Wiss., pp. 307–14.

Bernatchez, L. and Dodson, J.J. (1985) Influence of temperature and current speed on the swimming capacity of Lake whitefish (*Coregonus clupeaformis*) and cisco (*C. artedii*). *Can. J. Fish. Aquat. Sci.*, **42**, 1522–9.

Bernstein, M.H., Thomas, S.P. and Schmidt-Nielsen, K. (1973) Power input during flight of the fish crow, *Corvus ossifragus*. *J. exp. Biol.*, **58**, 401–10.

Blake, R.W. (1976) On seahorse locomotion. *J. Mar. Biol. Assoc. U.K.*, **56**, 939–49.

Blake, R.W. (1977) On ostraciiform locomotion. *J. Mar. Biol. Assoc. U.K.*, **57**, 1047–55.

Blake, R.W. (1978) On balistiform locomotion. *J. Mar. Biol. Assoc. U.K.*, **58**, 73–80.

Blake, R.W. (1983a) Swimming in the electric eels and knifefishes. *Can. J. Zool.*, **61**, 1432–41.

Blake, R.W. (1983b) Functional design and burst-and-coast swimming in fishes. *Can. J. Zool.*, **61**, 2491–4.

Blickhan, R. (1992) *Biomechanik der axialen aquatischen und der pedalen terrestrischen Lokomotion*, Habilitationsschrift, Universität des Saarlandes, 191 pp.

Blight, A.R. (1976) Undulatory swimming with and without waves of contraction. *Nature, Lond.*, **264**, 352–4.

Blight, A.R. (1977) The muscular control of vertebrate swimming movements. *Biol. Rev.*, **52**, 181–218.

Boddeke, R., Slijper, E.J. and Stelt, A. van der (1959) Histological characteristics of the body musculature of fishes in connection with their mode of life. *Proc. K. Ned. Akad. Wet. Ser. C Biol. Med. Sci.*, **62**, 576–88.

Bone, Q. (1966) On the function of the two types of myotomal muscle fibre in elasmobranch fish. *J. Mar. Biol. Ass. U.K.*,**46**, 321–49.

Bone, Q. (1975) Muscular and energetic aspects of fish swimming, in *Swimming and Flying in Nature* (eds C.J. Brokaw and C. Brennen), Plenum Press, New York, pp. 493–528.

Bone, Q. (1978) Locomotor muscle, in *Fish Physiology*, Vol VII, *Locomotion*, (eds W.S. Hoar and D.J. Randall), Academic Press, London, pp. 361–424.

Bone, Q., Johnston, I.A., Pulsford, A. and Ryan, K.P. (1986) Contractile properties and ultrastructure of three types of muscle fibre in the dogfish myotome. *J. Muscle Res. Cell Motil.*, **7**, 47–56.

Borelli, A. (1680) *De motu animalium*, pars prima, Angeli Bernabo, Rome, 365 pp.

Breder, C.M. (1926) The locomotion of fishes. *Zoologica (N.Y.)*, **4**(5), 159–297.

Breder, C.M. and Edgerton, H.E. (1942) An analysis of the locomotion of the seahorse, *Hippocampus*, by means of high speed cinematography. *Ann. N.Y. Acad. Sci.*, **63**, 145–72.

Brett, J.R. (1964) The respiratory metabolism and swimming performance of young sockeye salmon. *J. Fish. Res. Bd Can.*, **21**, 1183–226.

Brett, J.R. (1967) Swimming performance of sockeye salmon (*Oncorhynchus nerka*) in relation to fatigue time and temperature. *J. Fish. Res. Bd Can.*, **24**, 1731–41.

Brett, J.R. (1972) The metabolic demand for oxygen in fish, particularly salmonids and a comparison with other vertebrates. *Respir. Physiol.*, **14**, 151–70.

Brett, J.R. (1982) The swimming speed of adult pink salmon (*Oncorhynchus gorbuscha*) at 20 °C and a comparison with sockeye salmon, *O. nerka*. *Can. Tech. Rep. Fish. Aquat. Sci.*, no.1143, (40 pp.)

Brett, J.R. and Groves, T.D.D. (1979) Physiological Energetics, in *Fish Physiology*, Vol. VIII, *Bioenergetics and Growth*, (eds W.S. Hoar, D.J. Randall and J.R. Brett) Academic Press, London, pp. 279–352.

Brett, J.R. and Sutherland, D.B. (1965) Respiratory metabolism of pumpkinseed (*Lepomis gibbosus*) in relation to swimming speed. *J. Fish. Res. Bd Can.*, **22**, 405–9.

Brill, R.W. and Dizon, A.E. (1979) Effect of temperature on isotonic twitch of white muscle and predicted speeds of skipjack tuna, *Katsuwonus pelamis*. *Env. Biol. Fishes*, **4**(3), 199–205.

Bullard, R.W. (1966) Physiological problems of space travel, in *Physiology*, 2nd edn (ed. E.E. Selkurt), Little, Brown, Boston, pp. 657–76.

Bullock, A.M. and Roberts, R.J. (1974) The dermatology of marine teleost fish. I. The normal integument. *Oceanogr. Mar. Biol. Ann. Rev.*, **13**, 383–411.

Butler, P.J., West, N.H. and Jones, D.R. (1977) Respiratory and cardiovascular responses of the pigeon to sustained, level flight in a wind tunnel. *J. exp. Biol.*, **71**, 7–26.

Butler, P.J., Wilsom, W.K. and Woakes, A.J. (1984) Respiratory, cardiovascular and metabolic adjustments during steady state swimming in the green turtle, *Chelonia mydas. J. Comp. Physiol.*, **154B**, 167–74.

Calow, P. (1985) Adaptive aspects of energy allocation, in *Fish Energetics: New Perspectives* (eds P. Tytler and P. Calow), Croom Helm, London, pp. 13–31.

Carey, F.G. and Teal, J.M. (1966) Heat conservation in tuna fish muscle. *Proc. Nat. Acad. Sci. U.S.A.*, **56**, 1464–9.

Carey, F.G. and Teal, J.M. (1969a) Regulation of body temperature by the bluefin tuna. *Comp. Biochem. Physiol.*, **28**, 205–13.

Carey, F.G. and Teal, J.M. (1969b) Mako and porbeagle: warmbodied sharks. *Comp. Biochem. Physiol.*, **28**, 199–204.

Carpenter, R.E. (1975) Flight metabolism of flying foxes, in *Swimming and Flying in Nature* (eds T.Y. Wu, C.J. Brokaw and C. Brennen), Plenum Press, New York, pp. 883–9.

Chevrel, R. (1913) Essai sur la morphologie et la physiologie du muscle latéral chez les poissons osseux. *Archs. Zool. exp. gen. Notes Rev.*, **52**, 473–607.

Craig, A.B. and Påsche, A. (1980) Respiratory physiology of freely diving harbor seals (*Phoca vitulina*). *Physiol. Zool.*, **53**, 419–32.

Curtin, N.A. and Woledge, R.C. (1988) Power output and force–velocity relationship of live fibres from white myotomal muscle of the dogfish, *Scyliorhinus canicula. J. exp. Biol.*, **140**, 187–97.

Dam, C.P. van (1987) Efficient characteristics of crescent-shaped wings and caudal fins. *Nature, Lond.* **325**, 435–7.

Damant, G.C.C. (1925) Locomotion of the sunfish. *Nature, Lond.*, **116**, 543.

Daniel, T.L. (1981) Fish mucus: *In situ* measurements of polymer drag reduction. *Biol. Bull. mar. biol. Lab., Woods Hole*, **160**, 376–82.

Daniel, T.L. (1988) Forward flapping flight from flexing fins. *Can. J. Zool.*, **66**, 630–8.

Davis, R.W., Williams, T.K. and Kooyman, G.L. (1985) Swimming metabolism of yearling and adult harbor seals, *Phoca vitulina. Physiol. Zool.*, **58**, 590–6.

Davison, W. and Goldspink, G. (1977) The effect of prolonged exercise on the lateral musculature of the brown trout (*Salmo trutta*). *J. exp. Biol.* **70**, 1–12.

Domenici, P. and Blake, R.W. (1991) The kinematics and performance of the escape response in the angelfish (*Pterophyllum eimekei*). *J. exp. Biol.*, **156**, 187–204.

Dorn, P., Johnson, L. and Darby, C. (1979) The swimming performance of nine species of common California inshore fishes. *Trans. Am. Fish. Soc.*, **108**, 366–72.

East, P. and Magnan, P. (1987) The effect of locomotor activity on the growth of brook charr, *Salvelinus fontinalis* Mitchell. *Can. J. Zool.*, **65**, 843–6.

Eckert, R., Randall, D. and Augustine, G. (1988) *Animal Physiology*, 3rd edn, W.H. Freeman, San Francisco, 683 pp.

Edman, K.A.P. (1979) The velocity of unloaded shortening and its relation to sarcomere length and isometric force in vertebrate muscle fibres. *J. Physiol., Lond.*, **291**, 143–59.

Edman, K.A.P., Mulieri, L.A. and Scubon-Mulieri, B. (1976) Non-hyperbolic force–velocity relationship in single muscle fibres. *Acta physiol. scand.*, **98**, 143–56.

Ellington, C.P., Machin, K.E. and Casey, T.M. (1990) Oxygen consumption of bumblebees in forward flight. *Nature, Lond.*, **347**, 472–3.

Farmer, G.J. and Beamish, F.W.H. (1969) Oxygen consumption of *Tilapia nilotica* in relation to swimming speed and salinity. *J. Fish. Res. Bd Can.*, **26**, 2807–21.

Fauré-Fremiet, E. (1938) Structure du derme téliforme chez les scombrides. *Archs. Anat. microsc. morph. exp.*, **34**, 219–30.

Fedak, M.A. and Seeherman, H.J. (1979) Reappraisal of energetics of locomotion shows identical costs in bipeds and quadrupeds including ostrich and horse. *Nature, Lond.*, **282**, 713–16.

Feldkamp, S.D. (1987) Swimming in the California sea lion: morphometrics, drag and energetics. *J. exp. Biol.*, **131**, 117–35.

Ford, E. (1937) Vertebral variation in teleost fishes. *J. Mar. Biol. Ass. U.K.*, **22**, 1–60.

Freadman, M.A. (1979) Role partitioning of swimming musculature of striped bass, *Morone saxatilis* Walbaum and bluefish, *Pomatomus saltatrix* L. *J. Fish Biol.*, **15**, 417–23.

Fry, F.E.J. and Hart, S.J. (1948) Cruising speed of goldfish in relation to water temperature. *J. Fish. Res. Bd Can.*, **7**, 169–75.

Fuiman, L.A. (1986) Burst-swimming performance of larval zebra danios and the effect of diel temperature fluctuations. *Trans. Am. Fish. Soc.*, **115**, 143–8.

Fuji, R. (1968) Fine structure of the collagenous lamella underlying the epidermis of the goby, *Chasmichtys gulosus*. *Annot. zool. Jpn.*, **41**(3), 95–106.

Geerlink, P.J. (1979) The anatomy of the pectoral fin in *Sarotherodon niloticus* Trewavas (Cichlidae). *Neth. J. Zool.*, **29**, 9–32.

Geerlink, P.J. (1983) Pectoral fin kinematics of *Coris formosa* (Teleostei, labridae). *Neth. J. Zool.*, **33**, 515–31.

Geerlink, P.J. (1987) The role of the pectoral fins in braking in mackerel, cod and saithe. *Neth. J. Zool.*, **37**, 81–104.

Geerlink, P.J. (1989) Pectoral fin morphology: a simple relation with movement pattern? *Neth. J. Zool.*, **39**, 166–93.

Geerlink, P.J. and Videler, J.J. (1974) Joints and muscles of the dorsal fin of *Tilapia nilotica* L. (Fam. Cichlidae). *Neth. J. Zool.*, **24**, 279–90.

Geerlink, P.J. and Videler, J.J. (1987) The relation between structure and bending properties of teleost fin rays. *Neth. J. Zool.*, **37**, 59–80.

Gleeson, T.T. (1979) Foraging and transport costs in the Galapagos marine iguana, *Amblyrhynchus cristatus*. *Physiol. Zool.*, **52**, 549–57.

Goldspink, G. (1977) Design of muscles in relation to locomotion. In *Mechanics and Energetics of Animal Locomotion* (eds R. McNeill Alexander and G. Goldspink), Chapman and Hall, London, 346 pp.

Gooding, R.M., Neill, W.H. and Dizon, A.E. (1981) Respiration rates and low-oxygen tolerance limits in skipjack tuna, *Katsuwonus pelamis*. *Fish. Bull.*, **79**, 31–48.

Goodrich, E.S. (1904) On the dermal fin rays of fishes, living and extinct. *Q. J. Microsc. Sci.*, **47**, 465–522.

Gordon, A.M.A., Huxley, A.F. and Julian, F.J. (1966) The variation in isometric tension with sarcomere length in vertebrate muscle fibres. *J. Physiol., Lond.*, **184**, 170–92.

Gordon, M.S., Chin, H.G. and Vojkovich, M. (1989) Energetics of swimming in fishes using different methods of locomotion: I. Labriform swimmers. *Fish Physiol. Biochem.*, **6**, 314–52.

Graaf, F. de, Raamsdonk, W. van, Asselt, E. van and Diegenbach, P.C. (1990a) Identification of motoneurons in the spinal cord of the zebrafish (*Brachydanio rerio*), with special reference to motoneurons that innervate intermediate muscle fibres. *Anat. Embryol.*, **182**, 93–102.

Graaf, F. de, Raamsdonk, W. van, Hasselbaink, H., Diegenbach, P.C., Mos, W.,

Smit-Onel, M.J., Asselt, E. van and Heuts, B. (1990b) Enzyme histochemistry of the spinal cord and the myotomal musculature in the teleost fish *Brachydanio rerio*. Effects of endurance training and prolonged reduced locomotory activity. *Z. mikrosk. -anat. Forsch.*, **104**, 593–606.

Graham, J.B., Koehrn, F.J. and Dickson, K.A. (1983) Distribution and relative proportions of red muscle in scombrid fishes: consequences of body size and relationships to locomotion and endothermy. *Can. J. Zool.*, **61**, 2087–96.

Gray, J. (1933a) Studies in animal locomotion: I. The movement of fish with special reference to the eel. *J. exp. Biol.*, **10**(4), 88–104.

Gray, J. (1933b) Studies in animal locomotion: II. The relationship between waves of muscular contraction and the propulsive mechanism of the eel. *J. exp. Biol.*, **10**(4), 386–90.

Gray, J. (1933c) Studies of animal locomotion: III. The propulsive mechanism of the whiting (*Gadus merlangus*). *J. exp. Biol.*, **10**(4), 391–400.

Greene, C.W. and Greene, C.H. (1913) The skeletal musculature of the king salmon. *Bull. U.S. Bur. Fish.*, **33**, 21–60.

Greer Walker, M. and Emerson, L. (1978) Sustained swimming speeds and myotomal muscle function in the trout, *Salmo gairdneri*. *J. Fish. Biol.*, **13**, 475–81.

Grillner, S. and Kashin, S. (1976) On the generation and performance of swimming in fish, in *Neural Control of Locomotion* (eds R.M. Herman, S. Grillner, P.S.G. Stein and D.G. Stuart), Plenum Press, New York, pp. 181–201.

Haas, H.J. (1962) Studies on mechanisms of joint and bone formation in the skeleton rays of fish fins. *Devl. Biol.*, **5**, 1–34.

Harder, W. (1975a) *Anatomy of Fishes*. Part I: Text, E. Schweizerbart'sche, Stuttgart, 612 pp.

Harder, W. (1975b) *Anatomy of Fishes*. Part II: Figures and plates, E. Schweizerbart'sche, Stuttgart, 132 pp.

Harper, D.G. and Blake, R.W. (1989) A critical analysis of the use of high-speed film to determine maximum accelerations of fish. *J. exp. Biol.*, **142**, 465–71.

Harper, D.G. and Blake, R.W. (1990) Fast-start performance of rainbow trout *Salmo gairdneri* and northern pike *Esox lucius*. *J. exp. Biol.*, **150**, 321–42.

Harper, D.G. and Blake, R.W. (1991) Prey capture and the fast-start performance of northern pike *Esox lucius*. *J.Exp.Biol.*, **155**, 175–92.

Harris, J.E. (1936) The role of the fins in the equilibrium of the swimming fish. I: Wind-tunnel tests on a model of *Mustelus canis* (Mitchell). *J. exp. Biol.*, **15**, 32–47.

Harris, J.E. (1937) The mechanical significance of the position and movements of the paired fins in the teleostei. *Papers from Torygas Laboratory*, **31**(7), 173–89.

He, P. (1986) Swimming performance of three species of marine fish and some aspects of swimming in fishing gears. PhD thesis, University of Aberdeen. (232 pp.)

He, P. and Wardle, C.S. (1986) Tilting behaviour of the Atlantic mackerel, *Scomber scombrus*, at low swimming speeds. *J. Fish. Biol.*, **29** (suppl. A), 223–32.

He, P. and Wardle, C.S. (1988) Endurance at intermediate swimming speeds of Atlantic mackerel, *Scomber scombrus* L., herring, *Clupea harengus* L., and saithe, *Pollachius virens* L. *J. Fish. Biol.*, **33**, 255–66.

Hebrank, M.R. (1980) Mechanical properties and locomotor functions of eel skin. *Biol. Bull. Mar. biol. Lab., Woods Hole.*, **158**, 58–68.

Hebrank, M.R. and Hebrank, J.H. (1986) The mechanics of fish skin: Lack of an 'external tendon' role in two teleosts. *Biol. Bull. mar. biol. Lab., Woods Hole*, **171**, 236–47.

Hertel, H. (1966) *Structure, Form and Movement*, Reinhold, New York, 251 pp.

Hess, F. (1983) Bending moments and muscle power in swimming fish. *Proc. 8th*

Australasian Fluid Mechanics Conference, University of New Castle, New South Wales, **Vol.2**, 12A1–3.

Hess, F. and Videler, J.J. (1984) Fast continuous swimming of saithe (*Pollachius virens*): a dynamic analysis of bending moments and muscle power. *J. exp. Biol.*, **109**, 229–51.

Hocking, B. (1953) On the intrinsic range and speed of flight of insects. *Trans. R. Ent. Soc. Lond.*, **104**, 223–345.

Hoerner, S.F. (1965) *Fluid–Dynamic Drag*, 2nd edn, Hoerner, Brick Town, N.J., 452 pp.

Hollister, G. (1936) Caudal skeleton of Bermuda shallow water fishes I. *Zoologica (N.Y.)*, **21**, 257–90.

Houssay, S.F. (1912) *Forme, Puissance et Stabilité des Poissons*, A. Hermann et fils, Paris, 372 pp.

Hudson, D.M. and Bernstein, M.H. (1983) Gas exchange and energy cost of flight in the white necked raven, *Corvus cryptoleucus*. *J. exp. Biol.*, **103**, 121–30.

Hunter, J.R. and Zweifel, J.R. (1971) Swimming speed, tail beat frequency, tail beat amplitude, and size in jack mackerel, *Trachyrus symmetricus*, and other fishes. *Fish. Bull.*, **69**, 253–67.

Ivlev, V.S. (1963) Energy consumption during the motion of shrimps. *Zool. Zh.*, **42**, 1465–71 (in Russian).

James, A.G. and Findlay, K.P. (1989) Effect of particle size and concentration on feeding behaviour, selectivity and rates of food ingestion by the cape anchovy *Engraulis capensis*. *Mar. Ecol. Progr. Ser.*, **50**, 275–94.

James, A.G. and Probyn, T. (1989) The relationship between respiration-rate, swimming-speed and feeding behavior in the cape anchovy, *Engraulis capensis* Gilchrist. *J. exp. mar. Biol. Ecol.*, **131**, 81–100.

Johnson, T.P. and Johnston, I.A. (1991) Power output of fish muscle fibres performing oscillatory work: effects of acute and seasonal temperature change. *J. exp. Biol.*, **157**, 409–23.

Johnston, I.A. (1981) Structure and function of fish muscles, in *Vertebrate Locomotion*, (Symposia of the Zoological Society of London) (ed. M.H. Day), Academic Press, London, pp. 71–113.

Johnston, I.A. and Altringham, J.D. (1991) Movement in water: constraints and adaptations, in *Biochemistry and Molecular Biology of Fishes* (eds P.W. Hochachka and T.P. Mommsen), Elsevier, Amsterdam, pp. 249–68.

Johnston, I.A. and Salamonski, J. (1984) Power output and force-velocity relationships of red and white muscle fibres from the pacific blue marlin (*Makaira nigricans*). *J. exp. Biol.*, **111**, 171–7.

Johnston, I.A., Fleming, J.D. and Crockford, T. (1990) Thermal acclimation and muscle contractile properties in cyprinid fish. *Am. J. Physiol.*, **259**, R231–6.

Josephson, R.K. (1985) Mechanical power output from striated muscle during cyclic contraction. *J. exp. Biol.*, **114**, 493–512.

Karman T. von, (1956) *Aerodynamik*, Interavia Verlag.

Karman T. von and Burgers, J.M. (1934) General aerodynamic theory, in *Aerodynamic theory* (ed. W.F. Durand), Springer, Leipzig.

Kashin, S.M. and Smolyaninov, V.V. (1969) Concerning the geometry of fish trunk muscles. *J. Ichthyol.* (Eng.Transl. Vopr. Ikhtiol.), **9**, 923–5.

Kashin, S.M., Feldman, A.G. and Orlovsky, G.N. (1979) Different modes of swimming of the carp, *Cyprinus carpio* L. *J. Fish. Biol.*, **14**, 403–5.

Kaufmann, R. (1990) Respiratory cost of swimming in larval and juvenile cyprinids. *J. exp. Biol.*, **150**, 343–66.

Kawamura, G., Darusu, C. and Yonemori, T. (1988) Group effect on swimming speed of fish. *Nippon Suisan Gakkaishi*, **54**, 1067.

Kilarski, W. and Kozlowska, M. (1987) Comparison of ultrastructural and morphometrical analysis of tonic, white and red muscle fibres in the myotome of teleost fish (*Noemacheilus barbatulus* L.). *Z. Mikrosk. Anat. Forsch.*, **101**, 636–48.

Kishinouye, K. (1923) Contributions to the comparative study of the so-called scombroid fishes. *L. Coll. Agric. Imper. Univ. Tokyo*, **8**, 293–475.

Kleiber, M. (1961) *The Fire of Life: an Introduction to Animal Energetics*, Wiley, New York.

Koch, F. and Wieser, W. (1983) Partitioning of energy in fish: can reduction of swimming activity compensate for the cost of production?. *J. exp. Biol.*, **107**, 141–6.

Kutty, M.N. (1969) Oxygen consumption in the mullet *Liza macrolepis* with special reference to swimming velocity. *Mar. Biol.*, **4**, 239–42.

Langfeld, K.S., Altringham, J.D. and Johnston, I.A. (1989) Temperature and the force–velocity relationship of live muscle fibres from the teleost *Myoxocephalus scorpius*. *J. exp. Biol.*, **144**, 437–48.

Lannergren, L., Lindblom, P. and Johansson, B. (1982) Contractile properties of two varieties of twitch muscle fibres in *Xenopus laevis*. *Acta physiol. scand.*, **114**, 523–35.

Lanzing, W.J.R. (1976) The fine structure of fins and finrays of *Tilapia mossambica* (Peters). *Cell Tissue Res.*, **173**, 349–56.

Lasiewski, R.C. (1963) Oxygen consumption of torpid, resting, active, and flying hummingbirds. *Physiol. Zool.*, **36**, 122–40.

Leeuwen, J.L. van, Lankheet, M.J.M., Akster, H.A. and Osse, J.W.M. (1990) Function of red axial muscles of carp (*Cyprinus carpio* L.): recruitment and normalized power output during swimming in different modes. *J. Zool., Lond.*, **220**, 123–45.

LeFebvre, E.A. (1964) The use of D2O18 for measuring energy metabolism in *Columba livia* at rest and in flight. *Auk*, **81**, 403–16.

Leonard, J.B. and Summers, R.G. (1976) The ultrastructure of the integument of the american eel, *Anguilla rostrata*. *Cell Tissue Res.*, **171**(1), 1–30.

Lighthill, M.J. (1960) Note on the swimming of slender fish. *J. Fluid Mech.*, **9**, 305–17.

Lighthill, M.J. (1969) Hydromechanics of aquatic animal locomotion. *A. Rev. Fluid Mech.*, **I**, 413–46.

Lighthill, M.J. (1970) Aquatic animal propulsion of high hydromechanical efficiency. *J. Fluid Mech.*, **44**, 265–301.

Lighthill, M.J. (1971) Large-amplitude elongated-body theory of fish locomotion. *Proc. R. Soc.*,**179B**, 125–38.

Lighthill, M.J. (1975) *Mathematical Biofluiddynamics*, Soc. Indus. Appl. Math. Philadelphia, 281 pp.

Lighthill, M.J. and Blake, R. (1990) Biofluiddynamics of balistiform and gymnotiform locomotion. Part I. Biological background and analysis of elongated-body theory. *J. Fluid Dynamics*, **212**, 183–207.

Lindsey, C.C. (1978) Form, function and locomotory habits in fish, in *Fish Physiology*, Vol. VII, *Locomotion* (eds W.S. Hoar and D.J. Randall), Academic Press, London, pp. 1–100.

Lucas, M.C., Priede, I.G., Armstrong, J.D., Gindy, A.N.Z. and Vera, L. de (1991) Direct measurements of metabolism, activity and feeding behaviour of pike, *Esox lucius* L., in the wild, by the use of heart rate telemetry. *J. Fish Biol.*, **39**, 325–45.

Lumley, J.L. (1969) Drag reduction by additives. *A. Rev. Fluid Mech.*, **3**, 367–84.

McCleave, J.D. (1980) Swimming performance of european eel (*Anguilla anguilla* (L.)) elvers. *J. Fish Biol.*, **16**, 445–52.

McCutchen, C.W. (1970) The trout tail fin: a self cambering hydrofoil. *J. Biomech.*, **3**, 271–81.

McCutchen, C.W. (1976) Flow visualization with stereo shadowgraphs of stratified fluid. *J. exp. Biol.*, **65**, 11–20.

McCutchen, C.W. (1977) Froude propulsive efficiency of a small fish, measured by wake visualisation, in *Scale Effects in Animal Locomotion* (ed. T.J. Pedley), Academic Press, London, pp. 339–63.

McVean, A.R. and Montgomery, J.C. (1987) Temperature compensation in myotomal muscle: Antarctic versus temperate fish. *Env. Biol. Fishes*, **19**(1), 27–33.

Machin, K.E. and Pringle, J.W.S. (1960) The physiology of insect fibrillar muscle: III: The effects of sinusoidal changes of length on a beetle flight muscle. *Proc. R. Soc. Lond.*, **152B**, 204–25.

Magnuson, J.J. (1970) Hydrostatic equilibrium of *Euthymnus affinis*, a pelagic teleost without a gas bladder. *Copeia*, **1970**, 56–85.

Magnuson, J.J. (1973) Comparative study of adaptations for continuous swimming and hydrostatic equilibrium of scombroid and xiphoid fishes. *Fish. Bull.*, **71**, 337–56.

Marey, E.J. (1895) *Movement*, Heinemann.

Masman, D. and Klaassen, M. (1987) Energy expenditure during free flight in trained and free-living Eurasian kestrels (*Falco tinnunculus*). *Auk*, **104**, 603–16.

Mittal, A.K. and Banerjee, T.K. (1974) Structure and keratinization of the skin of a fresh water teleost *Notopterus notopterus* (Notopteridae, *Pisces*). *J. Zool. Lond.*, **174**, 341–55.

Motta, P.J. (1977) Anatomy and functional morphology of dermal collagen fibres in sharks. *Copeia*, **1977**, 454–64.

Müller, U.K. (1992) Dehnungen und Spannungen in der Haut von Regenbogenforellen (*Oncorhynchus mykiss*): Messungen *in vivo* und *in vitro*. Diplomarbeit, Universität Bielefeld. (108 pp.)

Müller, U.K., Blickhan, R. and Alexander, R.McN. (1991) New functional aspects of fish skin during locomotion. *J. Mar. Biol. Ass. U.K.*, **71**, 738.

Nachtigall, W., Rothe, U., Feller, P. and Jungmann, R. (1989) Flight of the honeybee III. Flight metabolic power calculated from gas analysis, thermoregulation and fuel consumption. *J. Comp. Physiol.*, **158**, 729–37.

Nadel, E.R., Holmer, I., Bergh, U., Astrand,P.O. and Stolwijk, J.A. (1974) Energy exchanges of swimming man. *J. Appl. Physiol.*, **36**, 465–71.

Nagy, K.A., Siegfried, W.R. and Wilson, R.P.(1984) Energy utilization by free-ranging jackass penguins, *Sphenicus demersus*. *Ecology*, **65**, 1648–55.

Nelson, J.S. (1984) *Fishes of the World*, 2nd edn, John Wiley & Sons, 523 pp.

Nishi, S. (1938) Muskelsystem II. Muskeln des Rumpfes. *Bolks Handb. Vgl. Anat. Wirbeltiere*, **5**, 351–446.

Nolet, B.A., Butler, P.J., Masman,D. and Woakes, A.J. (1992) Estimation of daily energy expenditure from heart rate and doubly labelled water in exercising geese. *Physiol. Zool.*, **65**, 1188–216.

Norberg, U.M. (1990) *Vertebrate Flight*, Zoophysiology **27**, Springer-Verlag, Berlin, 291 pp.

Norman, J.R. (1960) *A History of Fishes*, Ernest Benn, London, 463 pp.

Nursall, J.R. (1956) The lateral musculature and the swimming of fish. *Proc. R. Soc.*, **126B**, 127–43.

Nursall, J.R. (1963) The hypurapophysis, an important element of the caudal skeleton. *Copeia*, **1963**, 458–8.

O'Dor, R.K. (1982) Respiratory metabolism and swimming performance of the squid, *Loligo opalescens*. *Can. J. Fish. Aquat. Sci.*, **39**, 580–87.

Parsons, G.R. (1990) Metabolism and swimming efficiency of the bonnethead shark *Sphyrna tiburo*. *Mar. Biol.*, **104**, 363–7.

Peters, R.H. (1983) *The Ecological Implications of Body Size*, Cambridge University Press, Cambridge, 329 pp.

Pettigrew, J.B. (1873) *Animal Locomotion, or Walking, Swimming, and Flying with a Dissertation on Aeronautics*, Henry S. King, London.

Prandtl, L. and Tietjens, O.G. (1934) *Applied Hydro-and Aerodynamics* (transl. J.P. den Hartog) Dover (unabridged republication, 1957), 311 pp.

Prange, H.D. (1976) Energetics of swimming of a sea turtle. *J. exp. Biol.*, **64**, 1–12.

Prange, H.D. and Schmidt-Nielsen, K. (1970) The metabolic costs of swimming in ducks. *J. exp. Biol.*, **53**, 763–77.

Quinn, T.P. (1988a) Estimated swimming speeds of migrating sockeye salmon. *Can. J. Zool.*, **66**, 2160–63.

Raamsdonk, W. van, Kronnie G. te, Pool, C.W. and Laarse W. van de (1980) An immune histochemical and enzymic characterization of the muscle fibres in myotomal muscle of the teleost *Brachydanio rerio*, Hamilton Buchanan. *Acta histochem.*, **67**, 200–216.

Raamsdonk, W. van, Pool, C.W., Mijzen, P., Mos, W., Stelt, A. van der (1977) On the relation between movements and the shape of the somites in early embryos of the teleost *Brachydanio rerio*. *Bijdr. Dierk.*, **46**, 261–74.

Raso, D.S. (1991) A study of the peripheral innervation and muscle fibre types of *Ictalurus nebulosus* (Lesueur) and *Ictalurus punctatus* (Rafinesque). *J. Fish Biol.*, **39**, 409–19.

Raven, H.C. (1939) On the anatomy and evolution of the locomotor apparatus of the nipple tailed ocean sunfish. *Bull. Am. Mus. nat. Hist.*, **76**, 143–50.

Regnard, P. (1893) Sur un dispositif qui permet de mesurer la vitesse de translation d'un poisson se mouvant dans l'eau. *C.r. Séanc. Soc. Biol.*, **9**(5), 81–3.

Robertson, D.R. and Warner, R.R. (1978) Sexual patterns in the labroid fishes of the Western Caribbean 2. The parrotfishes. *Smithsonian Contr. Zool.*, **255**, 1–26.

Robinson, C.J. and Pitcher, T.J. (1989) The influence of hunger and ration level on shoal density, polarization and swimming speed of herring, *Clupea harengus* L. *J. Fish Biol.*, **34**, 631–3.

Rome, L.C. (1990) The influence of temperature on muscle recruitment and function in vivo. *Am. J. Physiol.*, **259**, R210–22.

Rome, L.C. and Sosnicki, A.A. (1991) Myofilament overlap in swimming carp II. Sarcomere length changes during swimming. *Am. J. Physiol.*, **260**, C289–96.

Rome, L.C., Funke R.P., Alexander, R.McN., Lutz, G., Aldridge, H., Scott, F. and Freadman, M. (1988) Why animals have different muscle fibres. *Nature Lond.*, **335**, 824–7.

Rome, L.C., Funke, R.P. and Alexander, R.McN. (1990) The influence of temperature on muscle velocity and sustained performance in swimming carp. *J. exp. Biol.*, **154**, 163–78.

Rosen, M.W. (1959) *Waterflow about a Swimming Fish*, U.S. Naval Ordnance Test Station, TP 2298, China Lake, California, 96 pp.

Rosen, M.W. and Cornford, N.E. (1971) Fluid friction of fish slimes. *Nature, Lond.*, **234**, 49–51.

Rothe, H.J., Biesel, W. and Nachtigall, W. (1987) Pigeon flight in a windtunnel. II. Gas exchange and power requirements. *J. Comp. Physiol.*, **157B**, 99–109.

Schmidt-Nielsen, K. (1972b) Locomotion: energy cost of swimming, flying and running. *Science*, **177**, 222–8.

Schmidt-Nielsen, K. (1990) *Animal Physiology: Adaptation and Environment*, Cambridge University Press, Cambridge, 620 pp.

Sengbush, R. von and Meske, C. (1967) Auf dem Wege zum grätenlose Karpfen. *Züchter*, **37**, 271–4.

Shann, E.W. (1914) On the nature of the lateral muscle in teleostei. *Proc. zool. Soc. Lond.*, **22**, 319–37.

Shuleikin, V.V. (1928) *Aerodynamics of Flying Fish*, Izv. AN SSSR, Ser. 7, Otd. Mat.i Estest. Nauk, **6–7**, 573–582.

Simons, J.R. (1970) The direction of the thrust produced by heterocercal tails of two dissimilar elasmobranchs: the Port Jackson shark, *Heterodontus portusjacksoni* (Meyer), and the piked dogfish, *Squalus megalops* (Macleay). *J. exp. Biol.*, **52**, 95–107.

Smit, H., Amelink-Koutstaal, J.M., Vijverberg, J. and Vaupel-Klein, J.C. von (1971) Oxygen consumption and efficiency of swimming goldfish. *Comp. Biochem. Physiol.*, **39A**, 1–28.

Sosnicki, A.A., Loesser, K.E. and Rome, L.C. (1991) Myofilament overlap in swimming carp I. Myofilament lengths of red and white muscle. *Am. J. Physiol.*, **260**, C283–88.

Squire, J.M., Luther, P.K. and Morris, E.P. (1990) Organisation and properties of the striated muscle sarcomere, in *Molecular Mechanisms in Muscular Contraction*, (ed. J.M. Squire), Macmillan, London, pp. 1–48.

Stainsby, W.N., Gladden, L.B., Barclay, J.K. and Wilson, B.A. (1980) Exercise efficiency: validity of base line subtractions. *J. Appl. Physiol.*, **48**, 518–52.

Stanfield, R.I. (1967) *Flying Manual and Pilots' Guide*, Ziff-Davis.

Stelt, A. van der (1968) Spiermechanica en myotoombouw bij vissen. Thesis, University of Amsterdam. (94 pp.).

Stevens, E.D. and Neill, W.H. (1978) Body temperature relations of tunas, especially skipjack, in *Fish Physiology*, Vol VII, *Locomotion* (eds W.S. Hoar and D.J. Randall), Academic Press,London, 315–59.

Sumich, J.L. (1983) Swimming velocities, breathing patterns, and estimated costs of locomotion in migrating gray whales, *Eschrichtius robustus*. *Can. J. Zool.*, **61**, 647–52.

Symmons, S. (1979) Notochordal and elastic components of the axial skeleton of fishes and their functions in locomotion. *J. Zool., Lond.*, **189**, 157–206.

Taylor, E.B. and Foote, C.J. (1991) Critical swimming velocities of juvenile sockeye salmon and kokanee, the anadromous and non-anadromous forms of *Oncorhynchus nerka* (Walbaum). *J. Fish Biol.*, **38**, 407–19.

Taylor, E.B. and McPhail, J.D. (1986) Prolonged and burst swimming in anadromous and freshwater threespine sticklebacks. *Can. J. Zool.*, **64**, 416–20.

Taylor, J.W.R. (1968) *Jane's all the World's Aircraft*. McGraw Hill, New York.

Thillart G. van den, (1986) Energy metabolism of swimming trout (*Salmo gairdneri*) Oxidation rates of palmitate, glucose, lactate, alanine, leucine and glutamate. *J. comp. Physiol*, **156B**, 511–20.

Thomas, S.P. (1975) Metabolism during flight in two species of bats, *Phyllostomus hastatus* and *Pteropus gouldii*. *J. exp. Biol.*, **63**, 273–93.

Thomson, K.S. (1976) On the heterocercal tail in sharks. *Paleobiology*, **2**, 19–38.

Torre-Bueno, J.R. and Larochelle, J. (1978) The metabolic cost of flight in unrestrained birds. *J. exp. Biol.*, **75**, 223–9.

Tsukamoto, K., Kajihara, T. and Nishiwaki, M. (1975) Swimming ability of fish. *Bull. Jap. Soc. scient. Fish.*, **41**, 167–74.

Tucker, V.A. (1968) Respiratory exchange and evaporative water loss in the flying budgerigar. *J. exp. Biol.*, **48**, 67–87.

Tucker, V.A. (1969) The energetics of bird flight. *Scient. Am.*, **220**, 70–8.

Tucker, V.A. (1970) Energetic cost of locomotion in animals. *Comp. Biochem. Physiol.*, **34**, 841–6.

Tucker, V.A. (1972) Metabolism during flight in the laughing gull, *Larus atricilla. Am. J. Physiol.*, **222**, 237–45.

Tucker, V.A. (1975) The energetic cost of moving about. *Am. Scient.*, **63**, 413–9.

Tytler, P. (1969) Relationship between oxygen consumption and swimming speed in the haddock, *Melanogrammus aeglefinus. Nature, Lond*, **221**, 274–5.

Videler, J.J. (1975) On the interrelationships between morphology and movement in the tail of the cichlid fish *Tilapia nilotica* L. *Neth. J. Zool.*, **25**, 144–94.

Videler, J.J. (1977) Mechanical properties of fish tail joints. *Fortschr. Zool.*, **24**(2–3), 183–94.

Videler, J.J. (1981) Swimming movements, body structure and propulsion in cod *Gadus morhua*, in *Vertebrate Locomotion*, (Symposia of the Zoological Society of London)(ed. M.H. Day), Academic Press, London, pp. 1–27.

Videler, J.J. (1985) Fish swimming movements: a study of one element of behaviour. *Neth. J. Zool.*, **35**, 170–85.

Videler, J.J. (1988) Sleep under sand cover of the labroid fish *Coris julis*, in *Sleep '86*, (eds W.P. Koella, F. Ob l, H. Schultz and P. Visser), Gustav Fischer, Stuttgart, pp. 145–7.

Videler, J.J. (1992) Comparing the cost of flight: Aircraft designers can still learn from nature, in *Biona* (Bionik-Kongress, Wiesbaden 1992) (ed. W. Nachtigall), Gustav Fischer, Stuttgart, pp. 53–72.

Videler, J.J. and Beukema, W.J. (1973) Apparatus for electromyographic recording in unanaesthetised free-swimming fish. *Med. Biol. Eng.*, **March**, 230–32.

Videler, J.J. and Hess, F. (1984) Fast continuous swimming of two pelagic predators, saithe (*Pollachius virens*) and mackerel (*Scomber scombrus*): a kinematic analysis. *J. exp. Biol.*, **109**, 209–28.

Videler, J.J. and Nolet, B.A. (1990) Costs of swimming measured at optimum speed: scale effects, differences between swimming styles, taxonomic groups and submerged and surface swimming. *Comp. Biochem. Physiol.*, **96**, 436–45.

Videler, J.J. and Wardle, C.S. (1978) New kinematic data from high speed cine film recordings of swimming cod (*Gadus morhua*). *Neth. J. Zool.*, **28**, 465–84.

Videler, J.J. and Wardle, C.S. (1991) Fish swimming stride by stride: speed limits and endurance. *Rev. Fish Biol. Fish.*, **1**, 23–40.

Videler, J.J. and Weihs, D. (1982) Energetic advantages of burst-and-coast swimming of fish at high speeds. *J. exp. Biol.*, **97**, 169–78.

Vleck, D., Gleeson, T.T. and Bartholomew, G.A. (1981) Oxygen consumption during swimming in the Galapagos marine iguanas and its ecological correlates. *J. comp. Physiol.*, **141B**, 531–6.

Wainwright, S.A. (1983) To bend a fish, in *Fish Biomechanics*, (eds P.W. Webb and D. Weihs), Praeger, New York, pp. 68–91.

Wainwright, S.A., Vosburgh, F. and Hebrank, J.H. (1978) Shark skin: Function in locomotion. *Science*, **202**, 747–9.

Wakeman, J.M. and Wohlschlag, D.E. (1982) Least-cost swimming speeds and transport costs in some pelagic estuarine fishes. *Fish. Res.* (Amst.), **1**, 117–27.

Walters, V. and Fierstine, H.L. (1964) Measurements of swimming speeds of yellowfin tuna and wahoo. *Nature, Lond.*, **202**, 208–9.

Wang, K. and Wright, J. (1988) Architecture of the sarcomere matrix of skeletal muscle: Immunoelectron microscopic evidence that suggests a set of parallel inextensible nebulin filaments anchored at the Z-line. *J. Cell Biol.*, **107**, 2199–212.

Wardle, C.S. (1975) Limit of fish swimming speed. *Nature, Lond.*, **255**, 725–7.

Wardle, C.S. (1977) Effects of size on the swimming speeds of fish, in *Scale Effects in Animal Locomotion* (ed. T.J. Pedley), Academic Press, London, pp. 299–313.

Wardle, C.S. (1980) Effect of temperature on the maximum swimming speed of fishes, in *The Environmental Physiology of Fishes* (ed. M.A. Ali), Plenum, New York, pp. 519–31.

Wardle, C.S. (1985) Swimming activity in marine fish, in *Physiological Adaptations of Marine Animals* (ed. M.S. Laverack), The Company of Biologists, Cambridge, pp. 521–40.

Wardle, C.S. and He, P. (1988) Burst swimming speeds of mackerel, *Scomber scombrus* L. *J. Fish Biol.*, **32**, 471–8.

Wardle, C.S. and Kanwisher, J.W. (1974) The significance of heart rate in free swimming cod, *Gadus morhua*: some observations with ultra-sonic tags. *Mar. Behav. Physiol.*, **2**, 311–24.

Wardle, C.S. and Reid, A. (1977) The application of large amplitude elongated body theory to measure swimming power in fish, in *Fisheries Mathematics* (ed. J.H. Steele), Academic Press, London, pp. 171–91.

Wardle, C.S. and Videler, J.J. (1980a) Fish swimming, in *Aspects of Animal Movement* (eds H.Y. Elder and E.R. Trueman), Cambridge University Press, Cambridge, pp. 125–50.

Wardle, C.S. and Videler, J.J. (1980b) How do fish break the speed limit? *Nature, Lond.*, **284**, 445–7.

Wardle, C.S. and Videler, J.J. (1993) The timimg of the EMG in the lateral myotomes of mackerel and saithe at different swimming speeds. *J. Fish Biol.* **42**, 347–59.

Wardle, C.S., Videler, J.J., Arimoto, T., Franco, J.M. and He, P. (1989) The muscle twitch and the maximum swimming speed of giant bluefin tuna, *Thunnus thynnus* L. *J. Fish Biol.*, **35**, 129–37.

Webb, P.W. (1971a) The swimming energetics of trout: I Thrust and power output at cruising speeds. *J. exp. Biol.*, **55**, 489–520.

Webb, P.W. (1971b) The swimming energetics of trout: II Oxygen consumption and swimming efficiency. *J. exp. Biol.*, **55**, 521–40.

Webb, P.W. (1973) Kinematics of pectoral fin propulsion in *Cymatogaster aggregata*. *J. exp. Biol.*, **59**, 697–710.

Webb, P.W. (1975) Hydrodynamics and energetics of fish propulsion. *Bull. Fish. Res. Bd Can.*, **190**, 1–159.

Webb, P.W. (1976) The effect of size on the fast-start performance of rainbow trout, *Salmo gairdneri*, and a consideration of piscivorous predators. *J. exp. Biol.*, **65**, 157–77.

Webb, P.W. (1978) Fast-start performance and body form in seven species of teleost fish. *J. exp. Biol.*, **74**, 211–26.

Webb, P.W. (1983) Speed, acceleration and manoeuvrability of two teleost fishes. *J. exp. Biol.*, **102**, 115–22.

Webb, P.W. (1984a) Bodyform, locomotion and foraging in aquatic vertebrates. *Am. Zool.*, **24**, 107–20.

Webb, P.W. (1984b) Form and function in fish swimming. *Scient. Am.*, **251**, 58–68.

Webb, P.W. (1986) Kinematics of Lake sturgeon, *Acipenser fulvescens*, at cruising speeds. *Can. J. Zool.*, **64**, 2137–41.

Webb, P.W. (1988) 'Steady' swimming kinematics of tiger musky, an Esociform accelerator, and rainbow trout, a generalist cruiser. *J. exp. Biol.*, **138**, 51–69.

Webb, P.W. (1993a) Is tilting behaviour at low swimming speeds unique to negatively buoyant fish? Observations on steelhead trout, *Oncorhynchus mykiss*, and blurgill, *Lepomis macrochirus*. *J. Fish Biol.*, in press.

Webb, P.W. (1993b) The effect of solid and porous channel walls on steady swimming of steelhead trout, *Oncorhynchus mykiss*. *J. exp. Biol.*, **178**, 97–208.

Webb, P.W. and Skadsen, J.M. (1979) Reduced skin mass: An adaptation for acceleration in some teleost fishes. *Can. J. Zool.*, **57**, 1570–75.

Webb, P.W., Kostecki, P.T. and Stevens, E.D. (1984) The effect of size and swimming speed on the locomotor kinematics of rainbow trout. *J. exp. Biol.*, **109**, 77–95.

Webb, P.W., Sims, D. and Schultz, W.W. (1991) The effects of air/water surface on the fast-start performance of rainbow trout (*Oncorhynchus mykiss*). *J. exp. Biol.*, **155**, 219–26.

Weihs, D. (1973a) The mechanism of rapid starting of a slender fish. *Biorheology*, **10**, 343–50.

Weihs, D. (1974) Energetic advantages of burst swimming of fish. *J. theor. Biol.*, **48**, 215–29.

Weihs, D. (1981) Body section variation in sharks: an adaptation for efficient swimming. *Copeia*, **1981**, 217–19.

Weihs, D. (1989) Design features and mechanics of axial locomotion in fish. *Am. Zool.*, **29**, 151–60.

Weis-Fogh, T. (1952) Fat combustion and metabolic rate of flying locusts (*Schistocerca gregaria Forskål). Phil. Trans. R. Soc.*, **237B**, 1–36.

Westerfield, M., McMurray, J.V. and Elsen, J.S. (1986) Identified motoneurons and their innervation of axial muscles in the zebrafish. *J. Neurosci.*, **6**, 2267–77.

Westerterp, K.R. and Drent, R.H. (1985) Energetic costs and energy-saving mechanisms in parental care of free-living passerine birds by the $D_2{}^{18}O$ method, in *Acta XVII Int. Ornith. Congr. Moscow* (eds V.D. Ilyichev and V.M. Gavrilov), Nauka, Moscow, pp. 392–8.

Wheeler, A. (1978) *Key to the Fishes of Northern Europe*, Frederick Warne, London, 380 pp.

Whoriskey, F.G. and Wootton, R.J. (1987) The swimming endurance of threspine sticklebacks, *Gasterosteus aculeatus* L., from the Afon Rheidol, Wales. *J. Fish. Biol.*, **30**, 335–9.

Willemse, J.J. (1959) The way in which flexures of the body are caused by muscular contractions. *Proc. K. Ned. Akad. Wet. Ser. C Biol. Med. Sci.*, **62**, 589–93.

Willemse, J.J. (1966) Functional anatomy of the myosepta in fishes. *Proc. K. Ned. Akad. Wet. Ser. C Biol. Med. Sci.*, **69**(1), 58–63.

Willemse, J.J. (1972) Arrangement of connective tissue fibres in the musculus lateralis of the spiny dogfish, *Squalus acanthias* L. (Chondrichthyes). *Z. Morphol. Tiere*, **72**, 231–44.

Williams, T.L., Grillner, S., Smoljaninov, V.V., Wallen, P., Kashin, S. and Rossignol, S. (1989) Locomotion in lamprey and trout: the relative timing of activation and movement. *J. exp. Biol.*, **143**, 559–66.

Williams, T.M. (1983) Locomotion in the North American mink, a semi aquatic mammal: I. Swimming energetics and body drag. *J. exp. Biol.*, **103**, 155–68.

Williams, T.M., Friedl, W.A., Fong, M.L. and Haun, J.E. (1991) Swimming in bottlenose dolphins (*Tursiops truncatus*): aerobic and anaerobic limits to performance. *J. mar. biol. Ass. U.K.*, **71**, 727–8.

Woakes, A.J. and Butler, P.J. (1983) Swimming and diving in tufted ducks, *Aythya fuligula*, with particular reference to heart rate and gas exchange. *J. exp. Biol.*, **107**, 311–29.

Woakes, A.J. and Butler, P.J. (1986) Respiratory, circulatory and metabolic adjustments during swimming in the tufted dusk, *Aythya fuligula. J. exp. Biol.*, **120**, 215–31.

Woledge, R.C., Curtin, N.A. and Homsher, E. (1985) *Energetic Aspects of Muscle Contraction* (Monographs of the Physiological Society), Academic Press, London, 357 pp.

Wolf, T.J., Schmid-Hempel, P., Ellington, C.P. and Stevenson, R.D. (1989) Physiological correlates of foraging efforts in honey-bees: oxygen consumption and nectar load. *Funct. Ecol.*, **3**, 417–24.

Author index

Species index

A. SCIENTIFIC NAMES

B. COMMON NAMES

Subject index